From Urbanization to Cities

The Politics of
Democratic Municipalism

From Urbanization to Cities

The Politics of
Democratic Municipalism

Murray Bookchin

Introduction by Sixtine van Outryve d'Ydewalle

Also by Murray Bookchin

From Urbanization to Cities: The Politics of Democratic Municipalism

ISBN 978-1-84935-438-7
E-ISBN: 978-1-84935-439-4
LCCN: 2021935964

AK Press   AK Press
370 Ryan Avenue #100   33 Tower Street
Chico, CA 95973   Edinburgh, EH6, 7BN
USA   Scotland
www.akpress.org   www.akuk.com
akpress@akpress.org   akuk@akpress.org

Please contact us to request the latest AK Press distribution catalog, which features
books, pamphlets, zines, and stylish apparel published and/or distributed by AK
Press. Alternatively, visit our websites for the complete catalog, latest news, and
secure ordering.

Cover design by John Yates, www.stealworks.com
Printed in the United States of America on acid-free, recycled paper

CONTENTS

Preface

This book, and particularly its title, has had a complicated life, not unlike the important archeological and anthropological discoveries that have occurred since it was first penned more than thirty years ago. Initially presented as a hardcover by Sierra Club Books in 1987 under the title *The Rise of Urbanization and the Decline of Citizenship*, it was later issued as a paperback in Canada in 1992 under the name *Urbanization Without Cities*, and finally by Cassell (now Bloomsbury) in 1995 under the current title: *From Urbanization to Cities*. In each case, my father searched for a more appropriate title, and with each edition, he made changes, some subtle and some major. The most significant occurred in the third edition in the closing chapter, "The New Municipal Agenda," where he sought to internationalize certain political questions previously discussed in the context of the United States by examining similar developments in Britain and the European continent. He also added a new appendix devoted to the nuts-and-bolts of confederal democracy that we have retained instead of the original appendix, "The Meaning of Confederalism," which now appears in his collection of essays on municipalism, *The Next Revolution: Popular Assemblies and the Promise of Direct Democracy*.

This new edition also undertakes to incorporate some advances in archeology and anthropology. Happily, they reinforce one of his important early arguments in the book: that the rise of early cities was not necessarily associated with agriculture or with economic exploitation, and that in many cases these early cities were egalitarian in nature. For his generous help with some of these new archeological interpretations on the subject of the Çatalhöyük people in chapter two, David Wengrow has my deepest gratitude. Any lingering errors reflect the many advances in archeology and anthropology since this book was written and still underway today.

Additionally, it is clear that terms widely used in the 1980s and 1990s like "citizen" have come to have different, often exclusive, or disparaging meanings today. I have tried, where possible in this new edition, to update or account for these changing interpretations, but it has not been possible to change the fact that this book was first written three decades ago. I hope that the reader will be understanding that my father's many uses of the word "citizen" and "citizenship" should be interpreted in the most liberatory sense—as including everyone living within a given community, not according to the exclusive definition imparted by nation-states, which he so abhorred. Even where there may be archaisms, the central argument of this text—that municipalities can, and must, become the loci of a rational, egalitarian, and ecologically stable society—still resonates. Any failings of age aside, this book remains a deeply researched contribution to the question of how we can reclaim a citizens' politics firmly rooted in the rich revolutionary history of popular assemblies—a project more important now than ever.

Debbie Bookchin,
New York, NY August 2021

Introduction
Sixtine van Outryve d'Ydewalle

"From Commercy, we call for the creation of popular committees throughout France, which function in regular general assemblies. Places where speech is liberated, where one dares to express oneself, to educate oneself, to help one another. If there must be delegates, it is at the level of each local Yellow Vests' popular committee, closer to the voice of the people. With imperative, recallable, and rotating mandates. With transparency. With trust."

—The Yellow Vests of Commercy, November 30, 2018

Ideas travel through time and place. Especially inspiring ones. From Murray Bookchin's typewriter in the United States of 1985, to a group of Yellow Vests occupying the Commercy city center in northeastern France some thirty years later, ideas about organizing at the local level in face-to-face popular assemblies, about educating oneself to debate and decide on public matters, about de-professionalizing politics through delegates with imperative and recallable mandates have become material and real.

The New Municipal Agenda Bookchin elaborated in *From Urbanization to Cities* has given direction to the deep democratic aspirations of people in struggles around the world. In particular, Bookchin's thought has been a resource for the Yellow Vests, a grassroots protest movement for political and economic reform, inspiring the Commercy Yellow Vests to call on their fellow protesters throughout France to organize in popular assemblies and reject representative government. In addition to popular assemblies at the city level, the Yellow Vests also created confederal egalitarian structures to enable collective decision making: the Assembly of Assemblies, as well as the Commune of communes—as Bookchin urges in his 1998 essay "A Politics for the Twenty-First Century."

Even after the Yellow Vest movement was brutally repressed at the national level, the struggle for direct democracy continued locally. Some Yellow Vests decided to create a citizens' assembly and run a list for the municipal elections. They voted to tie candidates' mandates to the decisions of the local popular assembly—one of the strategies Bookchin proposed to radically restructure local politics in order to prioritize direct democracy. And what started in Commercy has shown itself to be much more than a local phenomenon. The Yellow Vest movement is just one of the many international struggles that are planting the seed of what Bookchin calls "libertarian municipalism" in the minds of people who are discontented with the practice of representative government. From the surge of municipalist movements across Europe, to the popular assemblies growing in countries throughout the Americas, to the Kurdish-led democratic confederalism of Rojava in Northeast Syria, more and more communities are bringing to life the ideas Bookchin unearthed from his study of popular history so many decades ago.

That human beings possess an intrinsic tendency to organize democratically in popular assemblies at the level of the city is the essential message of *From Urbanization to Cities: The Politics of Democratic Municipalism*. Such a tendency has unfolded throughout the centuries, from the Neolithic to the present, passing through classical Athens, medieval towns, *Comuneros* in Early Modern Spain, New England town meetings, the Paris Commune of both 1792 and 1871, and the Russian, German, Spanish, and Hungarian revolutions. What Bookchin shows us in this important book is that, throughout history, there is an enduring legacy of communal popular assemblies as a form of self-government, and of the city as the arena in which to develop politics and citizenship, despite the rise of the nation-state. Especially in times of social unrest, the assembly form of democracy has been the preferred vehicle for the community to act on its future.

These popular assemblies answer a human aspiration: to make political decisions on a directly democratic face-to-face basis. Indeed, Bookchin concludes that it is not a given that history inexorably leads to the creation of the nation-state, its coercive apparatus, and its professional, representative government. Rather, Bookchin posits that there exists an unfolding tendency of human beings towards egalitarian and democratic institutions. The masterstroke of Bookchin in this book is to show us how the given state of affairs—the nation-state as the main political unit, representative government as the way to exercise power—is in no case inevitable, but is instead a constructed status quo designed to favor hierarchy. And if it is constructed, then it can be undone.

To counter the nation-state, Bookchin has proposed a clear vision of what the city ought to be, instead of what it is. For that purpose, he offers an historical account of the role cities played decades, centuries, even millennia ago. Arriving in the present, he offers us a sharp analysis of the rise of urbanization in the modern era, a process generated by the combination of the forces of the nation-state, capitalism, and industrialism. The modern city expands and explodes into a vast homogeneous and anonymous megalopolis, destroying social bonds and threatening the very integrity of city life.

But he also demonstrates how urbanization is not an inherent feature of cities, but a socially avoidable one, showing us that cities do not need to be ever-growing gigantic impersonal agglomerations dominating nature and ruling us rather than allowing us to rule ourselves. They can be, and in fact used to be, a place for the exercise of direct democracy—that is, a place for us to meet, debate, and decide what we want to do collectively, rather than leaving this task to the "professional" politicians.

Indeed, even during periods of intense urbanization, working class people have recolonized community life, aided by the organized labor movement, forming what Bookchin called the "underground communal world" of the industrial era. In these assemblies and clubs, workers, middle-class people, and farmers met and kept political life alive even in highly centralized nation-states through the exercise of communal citizenship. These civic movements, which formed the foundation of radical uprisings throughout the nineteenth and twentieth centuries, were the products of neighborhood and local life. When one looks at the Paris Commune of 1871, clubs and neighborhood assemblies were the basis of the political life of this revolution: they assembled to discuss important economic, political, and social matters and pushed for more radical measures; they held elected *communards* in check, as well as defended Paris on the barricades against the government army.

Bookchin aims to show us that the world we know today, a world of state domination, ecological destruction, and ever-expanding capitalism does not necessarily need to be. That it has been otherwise. That it *can* be otherwise. That we can create institutions to expand human tendencies towards solidarity, mutual aid, sharing, equality, cooperation, and civic life, rather than towards exploitation, individualism, accumulation, and competition. And if we want to ensure the survival of our species, we *must* take on this project.

Throughout his work, Bookchin offers us an arsenal of conceptual weapons to face and overcome current crises: ecological, social, migratory, and democratic. In other books, specifically *The Ecology of Freedom, The Modern Crisis,* and *The Philosophy of Social Ecology*, he develops the idea of social ecology, an interdisciplinary philosophy that is based on the assumption that human domination over nature stems from human domination over other human beings, and suggests that our relationship with each other and with the natural world can only be healed when we eliminate every form of hierarchy and domination and reclaim control over our everyday lives. To better understand how to do this, *From Urbanization to Cities* confronts the question of how to put the principles of social ecology into action in the political sphere, offering an emancipatory political philosophy, libertarian municipalism, that recovers the city as a place for the development of an egalitarian and democratic civic life, rather than for bureaucratic domination, capital expansion, and exploitation of nature.

The political philosophy of libertarian municipalism, later encapsulated by the term "Communalism" by Bookchin, asks anew two correlated questions: What is the main political unit for a people to govern itself? And how should public power be exercised? Bookchin answers these questions together, advocating for the municipality as the political unit to realize direct democracy, since it is the only place where the people can gather, deliberate, and directly make decisions together, rather than relying on representatives to exercise public power.

Such decisions are not limited to the political sphere, leaving the economic sphere unattended, as traditional theories of participatory or deliberative democracy too often tend to do. Bookchin asserts that the economic sphere must be subordinated to the political one, leaving the task to formulate economic policies regarding production, distribution, and consumption to the entire community through the popular assembly. A municipalized economy would give true meaning to Marx's maxim, "From each according to his ability and to each according to his needs."

For questions going beyond the scope of a single municipality, including economic ones, he proposes to organize confederal councils composed of delegates endowed with imperative and recallable mandates from their popular assemblies. These delegates would no longer be political professionals like our current representatives, but rather spokespersons conveying the decisions of their assemblies to the confederal council. In contrast with the nation-state, a confederal form of organizing a large territory might have avoided one of the greatest plagues of our era: fascism, and its ideology of national unity. With nationalism and fascism once more on the rise, understanding and unpacking the roots of the nation-state is a necessity if we are to fight it—an urgent project that Bookchin undertakes in this book. Thus, Bookchin not only analyzes how the nation-state was born, but also offers us an alternative to counter it by constructing a new way of living together.

Bookchin's political philosophy charts a way out of the democratic crisis driving individuals away from politics, precisely because it aims at reconstructing the city as a place for people to develop a civic life and take their destiny into their own hands. It also puts forward another definition of citizenship: one that is not based on the arbitrary borders of the nation-state, but rather on political participation in the self-government of the city. With a new definition of citizenship beyond borders, Bookchin's political philosophy offers the dynamite necessary to blow up the edifice of the nation-state. It also offers a reconstructive vision that protects immigrants from the limbo of apolitical existence, social exclusion, and economic exploitation. It makes them, and all of us, citizens in the most complete sense of the term: that of being active participants in the political life of the municipality.

Because it is the first public space we enter when we leave our private sphere, the only human-scaled entity, and also the first place where newcomers arrive, the city is the ideal locus in which people can become reconstituted from separate individuals into a community. By exploring how the city used to, and could again, be a democratic and unalienated arena where humans reach their full potential in reason, freedom, and creativity through discourse and collective decision making, Bookchin aims at recovering two concepts: politics and citizenship. To be a citizen, one has to live in a *polis*, a village, town, or neighborhood, to convene popular assemblies where everyone can participate in politics. Through a genealogy of the state, he invites us to demystify and dismantle the state and its product, statecraft—the domain of professional politics—in order to recover the true meaning of politics as the art of deciding collectively the course of action of the community. By distinguishing politics from statecraft, he proposes a form of political engagement that ceases to pretend to know people's will, but that rather allows them to actually form such a will.

For Bookchin, politics consist not only of the direct self-government of the city by its residents, but also in the educational process of shaping individuals who can actively take part in such a self-government and act in the public interest, a capacity anesthetized by centuries of representative government. Education, which Bookchin calls *paideia* in reference to its crucial role in Athenian democracy, is fostered by participation in politics, not blind obedience to state laws and policies made by the ruling class. Indeed, in our contemporary society, individuals are considered passive, incompetent, and uninterested when it comes to politics, a mistaken anthropological inference about human nature as a result of an oppressive political system. As a consequence, places to experience participatory political education are lacking. Nonetheless, they can be found in the assemblies that flourish around the world during all kinds of political struggles, including that of the Yellow Vests—assemblies that are working to counter centuries of state conditioning. Even if these assemblies do not yet constitute the libertarian municipalist system Bookchin calls for, they can still play this necessary educational role.

But to reach such a communalist society, where should we begin? While this book aims mainly at convincing us that politics have previously been—and could be again—organized differently than the current nation-state and representative government, Bookchin does not leave us with only *ideals* of a democratic community. Even though he refuses to establish a one-size-fits all blueprint model of municipalist politics to avoid having it transformed into an inflexible and rigid dogma, he fleshes out details of a potential strategy. And in later essays, collected in the book *The Next Revolution: Popular Assemblies and the Promise of Direct Democracy*, Bookchin expands on his theory of Communalism and proposes additional concrete paths to enacting it.

Bookchin is well aware that the municipality—even when built around strong institutions of self-government to engage its citizens in public life—is not sufficiently strong enough to challenge the state. The strategy he proposes is therefore to create a system of dual power—referring to the 1917 Russian Revolution, where the provisional government shared power with the *soviets*—between, on the one hand, the confederation of communes based on local self-government, and, on the other, the nation-state, where the two would compete for political legitimacy. This situation of tension between the two forces is seen as a necessity in Bookchin's dialectical thought. He insists that the confederation of communes and the state cannot coexist in the long term. Bookchin believes that the former could win and replace the state, insofar as the state would have been hollowed out of its legitimacy.

To dismantle the legitimacy of the state in favor of the confederation of self-governed communes, we need first to realize self-government at the level of the municipality. Bookchin proposes two strategies in order to do so: an extra-institutional and an institutional one. The extra-institutional strategy consists of creating radically new and alternative institutions by building extra-legal popular assemblies to manage communal affairs independently of the existing political system. The institutional strategy entails participating in existing legal municipal institutions, like city councils, by presenting candidates whose mandates would be tied to the popular assembly's will, in order to radically transform these institutions. The goal is for popular assemblies, making decisions via direct democracy, to inform all policy enacted by the city council. From a dual power perspective, consolidating the power of communal popular assemblies at the expense of the state through either of these strategies is necessary if we wish to build a confederation of communes, and create a communalist society.

Bookchin is attuned to offering a political philosophy that can be adapted, modified, extended, and interpreted according to local conditions. And this is exactly how people have put it into practice across the world. Seeking to answer their deeply democratic aspirations, movements take Bookchin's ideas as inspiration and adapt them to the nuances of their local struggles. In turn, these movements, such as the radical feminist, democratic, and ecological revolution in Rojava, inspire others to follow suit.

This new edition of *From Urbanization to Cities* is timed perfectly to allow the potentialities inherent in Bookchin's work to be developed to their fullest by these ever-growing democratic political movements around the world. This book aims to show these movements that they are part of a broader trend in history, that of the underground communal world. It also encourages them to discover and recover this history, to find inspiration for their own struggles in past successes and failures. Because it proposes an answer to the crisis of democracy without falling into the trap of the nation-state's closed identity; because it enables an egalitarian communal form of economic production and distribution based on the municipalization of the economy; because it offers a means of relating to nature that does not rest on endless exploitation and subsequent destruction, Bookchin's Communalism has the potential to inspire a whole generation of people in quest of solidarity, equality, and democracy. This book is an essential guide for that journey.

Prologue

This book is not another discussion of city planning, nor is it primarily a criticism of urban life, although the latter concern is an important theme in its contents. It is above all an attempt to formulate a new politics—specifically, a confederal municipalist politics as distinguished from centralized, nationalist forms of statecraft. As the title of the closing chapter of the book—"The New Municipal Agenda"—suggests, I wish to advance an argument for extending local citizen-oriented power through village, town, and city confederations at the expense, and ultimately the removal, of the nation-state.

At a time when even the nation-state has become an ambiguous phenomenon whose authority is often overridden by multinational corporations, the citizen, such as he or she is defined at this moment in the twenty-first century, is losing any sense of identity or power over everyday life. A vast corporate economic and political system—imbued with a life of its own—threatens to supplant the already diminishing control ordinary people have over their lives and their future. A human-made machine, so to speak, seems to be replacing the people who created it, one that seems to be self-directing as well as all-encompassing.

The problems that citizen disempowerment raise have been the subject of numerous books and articles. But even in the most sympathetic literature on the subject, the solutions offered have been adaptive for the most part. Most of the writers on the subject of cities and citizens have advanced various ways of working *within* the parameters established by the nation-state, as though its replacement with any other form of political organization were beyond the realm of possibility. Nearly all advocates of stronger "democracy" or of various forms of "civic republicanism," both of which are presumably meant to increase citizen participation, accept the nation-state and its bureaucratic apparatus as an unchallengeable given. We are normally told that society is much too complex or too global to do without some sort of state apparatus. Thus, according to the best-intentioned argument, a "minimal state" is needed, hopefully one contained by a strong civil society.

But if history from earliest recorded times to the present has demonstrated anything, it is the implacable fact that state power is corruptive. None of the most idealistic and principled revolutionary leaders of the past lived comfortably with the corruptive effects of state power. Either they succumbed to it, or they consciously tried to diffuse it. The retention of state power destroyed the moral integrity of the most dedicated socialists, communists, and anarchists who held it for a time. The English, French, Russian, and Spanish revolutions provide compelling evidence of the capacity of state power to corrupt—a capacity that can no longer be regarded as a moral truism but rather, given its unrelenting nature, must be seen as an existential fact. To pursue state power—or to "seize" it, to use the language of traditional radicalism—is to guarantee that it will persist as a form of elitist manipulation, expand, and be brutally exercised as an instrument against a popular democracy.

A libertarian municipalist or confederal municipalist politics advances the best approach against "seizures" of state power and its retention by an elite, by slowly trying to *accrete* power for municipalities—initially, by acquiring *moral* power for municipal assemblies, as I have indicated in the closing chapter of this book. Libertarian or confederal municipalism seeks to expand the democratic institutions that still linger on in any modern republican system by opening them to the widest public participation possible at any given time. Hence the slogan that I have advanced: "Democratize the Republic! Radicalize the Democracy!" It is not that state power is to be "seized"—and then never relinquished—but that popular power is to be expanded until all power belongs to the institutions of a participatory democracy.

Unless we are to agree that the present competitive, accumulative, and agonistic society is the "end of history"—namely, the best social system humanity can achieve over the long course of its history—I submit that we must counterpose public power to the realities of oligarchical power. By this I mean that we must counterpose an emerging *political* power, based on a direct, popular citizens' democracy, to the *state* power exercised by various parliaments, ministries, and republics, not to speak of overtly authoritarian forms of coercive rule.

I should make clear that by the word *politics*, I do not mean "statecraft," which is what people ordinarily mean by politics and its practitioners, politicians. Nor do I regard the state simply as a form of administration; rather I view it as a professional apparatus with a monopoly of violence that is used by ruling classes to control meddlesome lower classes. Similarly, when I use the word *citizen*, I do not mean electoral "constituents," any more than I use the word *democracy* to denote a system of "representative government"—a term that becomes a blatant oxymoron when it is renamed a "representative democracy." In this book my use of the words *politics*, *citizen*, and *democracy* reflects the original and classical meaning these words had in the past, a meaning that is all but lost today.

If we are to recover politics, citizenship, and democracy we need not only to recover our concept of the city as a place in which we work and engage in everyday consociation; we also have to see the city as a public arena, in which we intermingle to discuss public affairs, such as ways of improving our lives as civic beings. Many of the squares in ancient, medieval, and Renaissance cities—often replicated throughout cities such as Florence and Venice on the neighborhood level—were places in which citizens congregated, argued about public business, and held open meetings to make decisions on the city's affairs. This process of civic articulation and rearticulation has not been sufficiently noted in the public mind, and it must be recognized if we are to have a new politics based on real citizenship.

This raises the question of what such a new politics means and why I anchor it in the recovery of city life rather than nation-states. The usual answers given to the question of what constitutes a city are often spatial and demographic in character, viewing the city as an area occupied by a closely interlocked, densely populated human community. My own definition of a city cannot be reduced to a single proposition. Like rationality, science, and technology, which I regard as defined by their own histories, I view the city as the *history* of the city. That is to say, I view the city as the cumulative development— or dialectic—of certain important social potentialities and of their phases of development, traditions, culture, and community features.

Defining the early city, I maintain, begins with the recognition of the city as a creative breach with humanity's essentially biological heritage, indeed the "metamorphosis" of that heritage into a new *social* form of evolution. The city was initially the arena *par excellence* for the transformation of human relationships from associations based on biological facts, such as kinship, to distinctly social facts, such as residential propinquity; for the emergence of increasingly secular forms of institutionalization; for often rapidly innovative cultural relations; and for universalizing economic activities that had been previously associated with age, gender, and ethnic divisions. In short, the city was the historic arena in which—and as a result of which— biological affinities were transformed into social affinities. It constituted the single most important factor that changed an ethnic folk into a body of secular citizens, and a parochial tribe into a universal *civitatis*, where, in time, the "stranger" or "outsider" could become a member of the community without having to satisfy any requirement of real or mythic blood ties to a common ancestor. Not only did political relationships replace kinship relationships; the notion of a shared *humanitas* replaced the exclusivity of the clan and tribe, whose biosocial claims to be "the People" had often excluded the "outsider" as an inorganic, exogenous, or even threatening "other."

Hence the city was historically the arena for the emergence of such universalistic concepts as "humanity"—and is potentially the arena for the reemergence of concepts of political self-regulation and citizenship, for the elaboration of social relations, and for the rise of a new civic culture. The steps from a consanguineous clan, tribe, and village to a *polis*, or political city; from blood brothers and sisters who were *born* into their social responsibilities to citizens who, in the best of circumstances, could freely decide on their civic responsibilities and determine their own affinities based on reason and secular interests—these steps constitute a meaningful definition of the city.

Cities, to be sure, can rise and fall. They can be parochial in their own ways. They can enjoy good fortune for a time or, owing to conflicts, totally disappear as the great mounds of long-buried cities in Meso-america and Mesopotamia attest. But once the city established firm roots in the history of social development, it acquired a conceptual reality that still persists, and it can still undergo many metamorphoses despite the disappearance or stagnation of individual cities. The city, in effect, has become a historic tradition—often a highly moral one—that tends to expand uniquely human traits and notions of freedom, and an idea of civic commonality that corrodes the parochial bonds of blood ties, gender distinctions, age status-groups, and ethnic exclusivity.

In *From Urbanization to Cities* I wish to explore the enormous value of cities—and towns—as remarkable human creations. I have tried to examine from a historical viewpoint the origins of cities, their role in shaping humanity as a highly unique and creative species, and the promise they offer as arenas for a new political and social dispensation. I have tried to examine how the city evolved, what forms it assumed over time, how it functioned as more than a mere market or center of production, and how citizens of a city interacted with one another to produce a form of what the great Roman thinker Cicero called "second nature"—that is, a *humanly* made "nature"—that existed in balance with the "first nature" we usually call the natural environment. Hence the citizens of a city are of no less concern to me than the city itself, for the city at its best became an ethical union of people, a moral as well as a socioeconomic community—not simply a dense collection of structures designed merely to provide goods and services for its anonymous residents.

I wish to redeem the city, to explore it not as a corrosive phenomenon but as a uniquely human, ethical, and ecological community whose members often lived in balance with nature and created institutional forms that sharpened human self-awareness, fostered rationality, created a secularized culture, enhanced individuality, and established institutional forms of freedom. At a time when the city's traditional functions have been grossly denatured by the rise of megalopolises, politicians, an all-embracing nation-state, and increasing authoritarian controls over the individual—all cheap electoral rhetoric about democracy to the contrary notwithstanding—it is vital that we search back into the past to find the elements of a true communalism, to see them in their less adulterated form, and to reformulate a synthesis of their best attributes for a more rational society than humanity has known at any time in the past.

Perhaps from such a synthesis we can gain a sense of hope, perspective, and the basis for concerted action. To let great civic attributes languish in the past while cybernetic and postmodernist "futurists" project the irrational present into the coming century would be to let the ideal of a rational society—what the great revolutionaries of the late eighteenth and nineteenth centuries called the "Commune of communes"—fade away from the memory of later generations.[1]

I should note that I am only too mindful of the defects of the past, few of which alas have been completely overcome. The Athenian *polis* was riddled and ultimately poisoned by slavery, patriarchalism, and imperialism. The finest of the medieval democratic cities were partly and eventually became completely oligarchical. The cities of the Renaissance and Enlightenment had strong authoritarian traits and were civic republics, except in those few cases—particularly the New England town during the American Revolution, and Paris during the Great French Revolution—where remarkable advances in democratic institutions flourished.

But taken together, there is a shared history that should not—and cannot—be ignored. Moreover, in exploring *politics*, as I use this word here, I am not concerned exclusively with a single, presumably exemplary city or its institutions. I am concerned with the Hellenic and medieval notion that a city must be an *ethical* union of citizens. And I am committed to an overarching ethical vision of what a city *ought* to be, not merely what it is at any given time.

The term *ought* is the stuff out of which ethics is usually made—with the difference that in my view the "ought" is not a formal or arbitrary regulative credo but the product of reasoning, of an unfolding rational process elicited or derived eductively from the potentialities of humanity to develop, however falteringly, mature, self-conscious, free, and ecological communities. I call this integration of the best in first or "biological" nature and second or "social" nature an emergently new "third" or *free* nature—that is, an ethical, humanly scaled community that establishes a creative interaction with its natural environment. Here human beings, consciously responding to a sense of obligation to the ecological integrity of the planet, bring their uniquely human rational, communicative, richly social, imaginative, and aesthetic capacities to the service of the nonhuman world as well as the human.[2]

This ethics of complementarity, as I call it, would be both a culmination of eons of natural evolution—once it guided human behavior in the cities of an ecological society—and a culminating point in the development of reason itself: a condition in which rational goals could be established by those living in new, ecologically oriented networks of cities, citizens in the sense of truly rational beings. For citizenship, too, is a process—as the Greeks so brilliantly saw—a process involving the social and self-formation of people into active participants in the management of their communities.

Not only does this book try to provide a theoretical framework for a new politics; it also advances a self-conscious practice in which confederal municipalists can engage in local electoral activity.

Minimally, the goal of such a practice would be to alter city and town charters where possible to enlarge civic democracy and to establish grassroots structures that may be obliged initially to rely on the moral authority of a new citizenry to countervail and hopefully supplant the growing power of the nation state over public life.

As I have stressed, none of the historical examples I present in the pages that follow represents a model or a "paradigm" or ideal image of what could or should be achieved in the future. All of these examples, which are cited for what they innovated rather than for what they constituted at any given time, were significantly tainted by shortcomings—notably, class divisions and antagonisms, and the exclusion of women and, often, the propertyless from public activity.

What they do demonstrate is the compelling fact that directly democratic institutions in fairly complex cities for their time, such as Paris in the eighteenth century, *could* emerge and function with extraordinary effectiveness until they were physically suppressed by oligarchies and authoritarian regimes.[3] They failed not because their popular institutions were too tenuous or utopian but because their leaders failed to organize resolute movements in the defense of civic liberties. Indeed, important leaders of direct democracies usually acted as soloists, appealing to a formless mass of supporters who had only a limited knowledge of the conflicts that surrounded them and whose ideas were too inchoate to make use of the possibilities available to them.

Hence the considerable attention I have given to the need for education in citizenship or *paideia*, and to the need to develop a general social consciousness—a theoretical coherence—as part of any municipalist struggle. In large part, this theoretical understanding includes a recognition that the achievement of a "Commune of communes," to which socialists and anarchists alike have long aspired, requires a completely uncompromising politics. It rests on the notion of creating a fundamental *duality of power* in which increasingly independent and confederated municipalities emerge in flat opposition to the centralized nation-state. As I shall emphasize again in the book, whatever power confederated municipalities gain can be acquired only at the expense of the nation-state, and whatever power the nation-state gains is acquired only at the expense of municipal independence.

In the force-field that exists between the two, either the municipalities and their confederations will increase their power by diminishing the power of the nation-state, or the nation-state will increase its power by diminishing the authority of the municipalities and their confederations. Thus, for a municipalist movement to run candidates for state, provincial, or national office would be absurd, in fact, a subversion of its very claim to seek a grassroots or participatory democracy, if only because any office beyond the municipal level is, almost by definition, a form of representation rather than participation.

Even more significantly, it would ignore the crucial fact that even as they run candidates *for* local offices, confederal municipalists are also running them *against* state, provincial, and national offices and institutions. The demand for municipal confederations is simultaneously a demand for opposition to the nation-state in all its forms, and to the illusion that the control of state, provincial, or national legislative bodies at the "top" is a precondition for the attainment of local power at the "bottom."

Not only would a drive for state, provincial, and national office relax the tension between the "top," the realm of statecraft, and the "bottom," the realm of an authentic politics; it would diminish the *educational* function of politics at the "bottom," which alone can become the realm of a new politics. Far from reaching greater numbers of people by running candidates for the summits of state power, such campaigns would confuse the distinction between politics and statecraft, between the participatory and the representative, between the confederal and the national. As such, the tension between these two opposing spheres of activity would be relaxed; the truly democratic nature of political education, which is based on face-to-face discourse between neighbors and citizens, would be replaced by Internet-mediated relationships; and the moral and educational thrust of a communalist or confederal municipalist approach would be lost as a heavy mist beclouded the distinction between politics and statecraft. A "Commune of communes" is not a "Republic of communes" or a "Commonwealth of communes"; indeed, as a confederation of municipalities, it stands uncompromisingly opposed to any specious attempts to reduce a confederation or commune to a republic or a commonwealth.

The distinction between politics and statecraft must be maintained all the more to assure that no pragmatic exigencies or parliamentary strategies—even if only to propagandize on a national scale—are invoked for seemingly confederal-municipalist ends. Indeed, the most effective impact of municipalist education comes precisely from the fact that it is municipal—that is to say, that it can be conducted only on a person-to-person basis and that its scope can be extended only by a movement that tries to reach every municipality in a region or nation. It is this kind of education that makes for trust, personal interaction, and face-to-face discussion, and that fosters the development of a face-to-face democracy. Its authentic starting point is the small study group, block clubs, neighborhood-based media, and personal discourse.

From Urbanization to Cities thus advances an appeal—for a new theoretical framework in which to develop a new politics (in this term's Hellenic meaning rather than in the parliamentary meaning imparted to it by the nation-state). It also advances an appeal for a self-conscious *practice* in which confederal municipalists engage in local electoral activity to alter city and town charters, restructure civic institutions to provide the public sphere for direct democracy, and bring the means of production under *citizen* control—not under the particularistic forms of "workers' control" that tend to degenerate into a form of collectivistic capitalism, or under forms of nationalized production, which enhance the state's authority with greater economic power.

The city is here to stay. Indeed, it has been a crucial part of human history and a factor in the making of the human mind for millennia. Can we afford to ignore it? Must we accept it as it is—as an entity that faces obliteration by a sprawling urbanization that threatens the countryside as well? Or can we give the city a new meaning, a new politics, a new sense of direction—and, also, provide new ideals of citizenship, many of which were in fact attained in great part during past times?

My writing of this book stemmed from my conviction as a social ecologist that there is a pressing need to view the city—more generally, the municipality, if we are to include towns as well—as an ecological enterprise, not merely as a logistical or structural one. It should be conceived in terms that explore how notions of domination and the historical development of hierarchy have led to the social, as well as natural, problems we face today. It is my hope that the reader will find in these pages not only a deeper understanding of the historical potential for freedom represented by the city, but a programmatic agenda for the creation of a new politics that combines the high ideals of a participatory citizenship with the recognition of what the city or town can be in a rational, free, and ecological society.

Murray Bookchin
Institute for Social Ecology
Burlington, Vermont
February 1995

Chapter One
Urbanization against Cities

The title of this chapter has been deliberately worded to create a paradox in the reader's mind. How, it may be fair to ask, can we speak of urbanization against cities? The two words, "urbanity" and "city," are usually taken to be synonyms. Indeed, as conventional wisdom would have it, a city is by definition an urban entity, and an urban entity, in turn, is certainly regarded as a city.

Yet I shall take great pains to show that they are in sharp contrast to each other—in fact, that they are bitter antagonists. My reasons for making such an unorthodox distinction between urbanization and citification are not intended to be semantic word juggling. The contradiction forms the very rationale for writing this book. "Urbanization against cities" is meant to focus as sharply as possible on a human and ecological crisis so deep-seated we are hardly aware of its existence, much less its grave impact on social freedom and personal autonomy. I refer to the historic decline of the city as an authentic arena of political life (that once lived in some balance with the natural world) and, perhaps no less significantly, the decline of the very notion of citizenship.

So sweeping a statement obviously requires some clarification at the outset of our discussion. According to most social theorists, the traditional "contradiction" created by the rise of urbanism has been the city's ages-old "conflict" with the countryside. History, we are commonly told, is filled with innumerable examples of the city's efforts to free itself from the trammels of agrarian parochialism. The city, it is emphasized, has always tried to assert its cosmopolitan culture and secular civic institutions over the narrow provincialism and constrictive kinship ties of the rural town and village. We take it for granted that from ancient to modern times the city and countryside have been at "war" with each other. Historically, this "war" is supposed to be embodied in the conflicting interests between the feudal lord and the urban merchant, the peasant food-cultivator and town-dwelling craftsperson, the land-based aristocrat and the citified capitalist, the farmer and the industrial worker.

That such conflicts have existed over the course of history and echo in modern society is true enough. We still retain ingrained visions of a cleansing and virtuous pastoral life that stands in moral contrast with the tainted and sinful world of the city. The contrast has been the subject of Biblical invectives, the theme of some of our most outstanding novels, and a centerpiece in the writings of many distinguished sociologists.

Taken as a whole, however, this Manichean drama can be a gross simplification of reality. It is certainly true that city and countryside have commonly viewed each other antagonistically in the past. And there have been long stretches of history when each tried to assert economic and political dominance over the other. But there have also been times when they existed in an almost exquisitely sensitive, creative, and ecological balance with each other. Some of the most admirable human adventures in culture, technics, and social freedom have occurred precisely in those periods when a complementary relationship between city and country, indeed between society and nature, successfully replaced the mutual rivalry for supremacy.

Today, it would seem that the city has finally achieved complete dominance over the countryside. Indeed, with the extension of suburbs into nearby open land on an unprecedented scale, the city seems to be literally engulfing the agrarian and natural worlds, absorbing adjacent towns and villages into sprawling metropolitan entities—a form of social cannibalism that could easily serve for our very definition of urbanization. All talk of metropolitan Babylon and Rome to the contrary, we have no comparable parallels in the past to urbanization on our present-day scale. Worse, the city seems to be replacing rural culture and all its rich traditional forms with the mass media and technocratic values we tend to associate with "city life." If all of this is true—as I frankly believe it is—I plan to introduce in this book a very discordant qualification. Contrary to most views on the subject, I plan to show that if we use words such as "city" and "country" meaningfully to describe this process of physical, cultural, and ecological urban cannibalism, the image of an all-devouring "city" that is engulfing a supine and helpless "country" is a sheer myth.

The truth is that the city *and* the country are under siege today—a siege that threatens humanity's very place in the natural environment. Both are being subverted by urbanization, a process that threatens to destroy their identities and their vast wealth of tradition and variety. Urbanization is engulfing not only the countryside; it is also engulfing the city. It is devouring not only town and village life based on the values, culture, and institutions nourished by agrarian relationships. It is devouring city life based on the values, culture, and institutions nourished by civic relationships. City space with its human propinquity, distinctive neighborhoods, and humanly scaled politics—like rural space, with its closeness to nature, its high sense of mutual aid, and its strong family relationships—is being absorbed by urbanization, with its smothering traits of anonymity, homogenization, and institutional gigantism. Whether or not the urbanization of both the city and the country is desirable is an issue that I shall earnestly explore. But I cannot emphasize too strongly that even if we think in the old terms of city versus country and the unique political contrasts such a time-honored imagery has nurtured, the conflict between city and country has largely become obsolete, Urbanization threatens to replace both contestants in this seemingly historic antagonism. It threatens to absorb them into a faceless urban world in which the words "city" and "country" will essentially become social, cultural, and political archaisms.

Perhaps our greatest difficulty in understanding urbanization and its grave impact on social and personal life today stems from our tendency to link it with our very naïve idea of the city. We are often satisfied to call any urban entity a "city" if it is demographically congested, structurally sizable, and, most significantly, populated by individuals whose work no longer deals directly with food cultivation. Urbanization, like citification, seems to meet these criteria so completely that we commonly identify the two, and distinguish them more as a matter of degree than of kind. Thus, we tend to regard a sprawling metropolitan area merely as an oversized city, or it may be an agglomeration of closely packed "cities" that Americans call "urban belts" and the British call "conurbations."

Granted that we are finding it increasingly difficult, on careful reflection, to regard urban belts and conurbations merely as cities. We uneasily sense that they are something more—if not something newer—than what previous generations called cities. What confuses us about such perplexing issues is that the people who live in such new urban entities are plainly engaged in city-type occupations and seem to follow citified lifeways. Increasingly removed from the natural world, metropolitan dwellers rarely, if ever, earn their livelihood as farmers, however fashionable urban and suburban gardening has become in recent years. They are employed in urban jobs, be they professional, managerial, service-oriented, craft-oriented, or the like. They live highly paced and culturally urbane lifeways that lock into mechanically fixed time slots—notably, the "nine-to-five" pattern—rather than follow agrarian cycles guided by seasonal change and dawn-to-dusk personal rhythms. Urban environments are highly synthetic rather than natural. Food is normally bought rather than grown. Dwellings tend to be concentrated rather than dispersed. Personal life is not open to the considerable public scrutiny we find in small towns or rooted in the strong kinship systems we find in the country. Urban culture is produced, packaged, and marketed as a segment of the city dweller's leisure time, not infused into the totality of daily life and hallowed by tradition as it is in the agrarian world. That country life is growing more and more like city life and losing its simplest natural attributes is a point that will become highly salient for the purposes of our discussion. For the present, what counts is that the aforementioned distinction still exists in the image of what we call urbanization (conceived as citification) as distinguished from ruralization.

Superficial as these continuities and contrasts may be, they begin to dissolve completely when we start to explore history for richer, fuller standards of what we mean by the word cities. I use the plural, cities, advisedly: Despite certain similarities, the differences between cities of the past also have considerable importance. History presents us with a wide range of unique and distinctive cities—the earliest in Sumer centering around temples; later ones, such as Babylon, around palaces; the more dynamic Greek democracies around civic squares that fostered citizen interaction; medieval and more recent ones, around a variety of different marketplaces. However diversified cities may have been in the past, our language gives them considerable prestige. The term civilization has its origin in *civitas*—a Latin word occasionally used for city—and denotes the cultural sophistication the western world has traditionally imparted to some kind of urbanism. Most of our utopian visions, whether heavenly or earthly, take the form of a city, a "New Jerusalem" to speak in sacred terms, or an idealized version of the Hellenic "city-state" to use secular language.

But here we abruptly encounter the limits of the term urbanization as a synonym for citification. Urbanization does not comfortably fit into an imagery of the city drawn from theocratic, monarchic, democratic, and economic communities peopled by craftsfolk and small merchants who were engaged in a natural economy. Our urban belts and conurbations are vast engines for operating huge corporate enterprises, industrial networks, distribution systems, and administrative mechanisms. Their facilities, like their towering buildings, stretch almost endlessly over the landscape until they begin to lack all definition and centrality. It is difficult to root them in a temple, palace, public square, or the small, intimate marketplace of craftsfolk and merchants. To say that they have any specific center that gives them civic identity is often so ill-fitting as to be absurd, even if one allows for centers that linger on from past eras when cities were still clearly delineable areas for human association.

What, then, do our premodern cities with their rich diversity of forms and functions have in common? We arrive at a basic characteristic of city life that is the result not only of human propinquity, structural size, occupations, and an urbane culture. What major cities of the past share—whatever their differences—are largely moral, often spiritual, attributes with deep roots in a natural environment that sharply distinguish them from the physical attributes we associate with urbanization. Cities of the past from their very beginnings were ultimately what I would call "communities of the heart"—moral associations that were nourished by a shared sense of ideological commitment and public concern. Civic ideology and concern centered around a strong belief in the good life for which the city provided the arena and catalytic agent. The good life by no means meant the affluent life, the life of personal pleasures and material security. More often than not, it meant a life of good*ness*, of virtue and probity. This sense of civic calling could assume a highly spiritual form such as we find in the Jews' reverence for Jerusalem or a highly ethical form as in the Greeks' admiration for Athens. Between these emotive and intellectual extremes, city dwellers of the past tended to form "social compacts" that were guided not only by material and defensive considerations but by loyalties to their cities that were shaped by richly textured ideological commitments.

Love of one's city, a deep and abiding sense of loyalty to its welfare, and an attempt to place these sentiments within a rich moral and ecological context, whether God-given or intellectual, clearly distinguishes the majority of cities of past eras from those of present ones. We have virtually no equivalent in the modern city of the Near Eastern sense of civic spirituality, the Greek feeling of political affiliation, the medieval endearment to communal fraternity, and the Renaissance love of urban pageantry that infused the otherwise disparate citizenry of the past. Such loyalties, with their moral underpinnings, may linger on in the residents of modern cities, but more as a fevered appetite for the material, cultural, and nervous stimulation of what we today designate as the good life than as a product of the ethical sturdiness, spiritual commitment, and sense of civic virtue that marked the citizenry of earlier eras. It also meant a love of the land, of place and natural setting that gave rise to a rich ecological sensibility and respect for the countryside.

To speak truthfully, our present-day relationship to the city usually takes the form of very pragmatic material requirements. A modern city, suburb, town, or, for that matter, a village is often evaluated in terms of the "municipal services" it offers to its residents. The traditional religious, cultural, ethical, and ecological features that once endeared citizens to their city, and its natural surroundings have dissolved into quantitative, often ethically neutral, criteria. The city is the first fund into which we make a series of social investments for the express purpose of receiving a number of distinctly material returns. We expect our persons and property to be protected, our shelters to be safeguarded, our garbage to be removed, our roads to be repaired, our environment to be physically and socially tidy—which is to say, spared from the invasion of "undesirable elements."

Doubtless we want our cities to be culturally stimulating, economically viable, and attractive in reputation. But such attainments are not necessarily a function of the city as such. They often depend upon the results of personal enterprise, the presence of individual types who reside in it by virtue of birth or choice. For a city to claim a famous son—and, more recently, a famous daughter—does not necessarily reflect well upon the city's reputation for producing gifted or renowned people but rather on the individual gifts of one or more of its residents. It does not provide us with evidence of the city's cultural milieu but of a specific person's gifts and biography that often involve a valiant effort to transcend a community's oppressive environment. Such was the case, for example, with James Joyce's relationship with Dublin, Oscar Wilde's relationship with London, or Cezanne's relationship with Paris.

Hence, urban "civilization" as we know it today is the erratic byproduct of a particular city, not its main and distinguishing consequence. Such a "civilization" emerges from the wayward activities of fairly privatized individuals or corporations rather than the innate traits of the municipality itself. A civic culture does not stem from the collective efforts of a unique and cohesive public, however unusual the city may be in terms of its regional location and its historic traditions. It stems from the personal activities of certain individuals who happen to occupy a residence within the city's confines, staff its shops and offices, work its industries, and, of course, produce its artistic artifacts. Urban "civilization," today, is not a characteristic civic phenomenon that emerges from a distinctive public and body politic; it is simply the exudate of free enterprise with its patina of "public service" and cultural charity. That mayors, corporate leaders, and philanthropists may vie with each other in celebrating the projects they initiate—projects that may range from concert halls and museums to airports and industrial parks—is simply evidence of the shallowness of what today is called "civic-mindedness." Rarely do these projects, which in any case are seldom free of vulgarity, nourish the city as a collectivity and arena for public activity. Like iridescent bubbles that rise, glisten, and burst, they form the surface of the city's often stagnant cultural life and social malaise.

In fact, like any marketplace, the modern city is the hectic center of a largely privatized interaction between anonymous buyers and sellers who are more involved in exchanging their wares than in forming socially and ethically meaningful associations. Cities today are typically measured more by their success as business enterprises than cultural foci. The ability of an urban entity to "balance its budget," to operate "efficiently," to "maximize" its service with minimal cost, all of these are regarded as the hallmark of municipal success. Corporate models form the ideal examples of urban models, and civic leaders take greater pride in their managerial skills than in their intellectual abilities.

This dominant entrepreneurial concept of the city has its precise counterpart in the dominant contemporary notion of citizenship. If we tend to view the city as our most immediate social "investment," we expect the city to give us adequate material "returns." We pay our taxes with a distinct expectation of the services they will buy. The greater the services for the money we pay, the more profitable it is to reside in a given city. Civic amenities are clearly measurable in terms of the number of schools, class sizes, parks, firehouses, transportation facilities, police, crime rates, parking spaces—indeed, in terms too numerous to inventory. When we "buy" into a residential area, we reconnoiter it primarily for these material and logistical amenities and secondarily, if at all, for the cultural stimulation and sense of community it provides.

Not surprisingly, the resident of most cities today tends to develop a very distinctive self-image. It is not the image of a citizen—a historically remarkable term that I have yet to describe—but that of a taxpayer. He or she does not have a sense of self appropriate to what we might call a public figure but rather that of a free-wheeling investor. Doubtless, taxpayers and investors often form many associations, but they ally with each other to advance or protect very specific interests. Like all sensible entrepreneurs who are involved in the business of residing in a city, they want a favorable return for what they pay, and, as the saying goes, "in numbers there is strength." Accordingly, all common adages to the contrary, they *can* fight City Hall—presumably, the place where the corporate board meets—and, depending as much upon their wealth as upon their numbers, they may succeed.

But beyond this economically secure interplay of conflicting interests and demands, the citizen *qua* taxpayer is not expected to get deeply involved in municipal affairs. Nor does the contemporary urban environment encourage him or her to do so. A "good citizen" is one who obeys the laws, pays taxes, votes ritualistically for pre-selected candidates, and "minds his or her own business." This notion of appropriate civic behavior is not merely a mutually shared, quietistic vision of modern-day citizenship; it is a political desideratum that, if violated, exposes more active taxpayers to charges of "meddling" at best and "vigilantism" at worst. Both taxpayers and municipal officials prudently acknowledge that the people of a city should be properly represented by efficient, specialized, and professional surrogates of "the public." Day-to-day power, however, resides precisely in the hands of these managerial surrogates, not in their "constituencies" who increasingly acquire the anonymity and face-lessness that the word "constituency" denotes. Like the traditional liberal concept that government is best when it governs least, so the contemporary liberal concept of citizenship seems to be that a "constituent" is best when he or she acts the least.

Such a concept of citizenship is fraught with grave psychological as well as political consequences. Individuals whose public lives barely transcend the social level of mere taxpayers tend to form very passive images of their personalities and of the natural environment around them. An increasingly disempowered citizen may well become a quietistic and highly retiring self. A major loss of social power tends to render a person less than human and thereby yields a loss of individuation itself. Such constituents live in a painfully contradictory world. On the one hand, society becomes an intensely problematic presence in their lives. The social realm is a potential source of war, of economic instability, of contending factions and ideologies that may reach directly into the most guarded niches of private life. These problems become particularly intimate when issues such as abortion, military conscription, sexual freedom, and earning capacity invade the individual's domestic realm.

On the other hand, while such issues rage around the "constituent," he or she is steadily divested of the power to act upon them. Indeed, the "constituent's" intellectual equipment to form an assured opinion is eroded by an ever-deepening sense of personal incompetence and public detachment. A preoccupation with trivia— the problems of shopping, fashion, personal appearance, career advancement, and entertainment in a thoroughly boring milieu— replaces the more heroic stance of a socially and environmentally involved body politic. We thus encounter a twofold development: a world in which growing social power preempts concerns that were once largely within the purview of the individual and the community, and the steady erosion of personal power and the individual's capacity for action. Within this paralyzing forcefield, the individual's self-identity begins to suffer a crucial decline. Self-recognition dissolves steadily into a grim lack of selfhood. Inaction becomes the only form of action with the result that the "constituent" retreats into an inwardness that lacks the substance to render one individually functional.

A world in which personality itself resembles the *tabula rasa* of an aimless society and a meaningless way of personal life would seem to raise more universal issues than the fate of the city and the citizen. But in many respects this universality expresses itself as a need for a larger perspective toward civic issues. The city is not only the individual's first social "investment"; it is his or her most intimate social environment. Owing to its vital immediacy, the city remains (as it has throughout history) the most direct arena in which the individual can act as a truly social being and from which he or she can attain the most immediate social solutions to the broader problems that beleaguer the privatized self. Insofar as the individual's self-definition as an empowered person and citizen is even possible today, the civic terrain on all its levels must be regained by its constituents and reconstituted in new ways to render people socially operational. Civic reempowerment of the citizen thus becomes a personal issue as well as a social one. It is equivalent to regaining one's private selfhood as well as one's public selfhood, one's personality as well as one's citizenship.

To attain such reempowerment and self-reconstitution has its presuppositions. So much of civic participation and civic-mindedness has been lost in the twentieth century, particularly its latter half, that we will have to probe deeply into the buried history of the city and of citizenship for reference points by which to understand where we stand in the swirl of urbanization that surrounds us.

We will want to know what the concepts of "city" and "citizenship" really mean—not simply as ideal definitions but as fecund ecological processes that reveal the growth of communities and the individuals who people them—indeed, that turn them into a genuine public sphere and a vital body politic. It is painfully characteristic of our present-day myopia that these very words, public sphere and body politic, have simply dropped out of our social vocabulary. When we use them at all, we rarely seem to understand what they mean or at least meant to earlier civilizations in which we claim to have our civic and social roots. They have been supplanted by such terms as "electorate" and, of course, "taxpayers" and "constituents"—administrative terms that denature the concepts of politics and community by definition.

It also behooves us to examine the kind of institutions cities have created to foster citizenship and public empowerment. A remarkable number of conflicting institutional issues clustered around the city as it meandered its way through different historical forms: decentralization versus centralization, direct democracy versus representative republicanism, assemblies of the people versus councils of deputies, recall and rotation of public officials versus lengthy tenure in office and professional fixity, popular management of social affairs versus bureaucratic control and manipulation. These issues have exploded repeatedly, from ancient times to the present, into bitter civic conflicts. They persist in our very midst under the rubric of "reform" movement to alter city charters and most recently as neighborhood movements to establish "grassroots" democracy.

We will want to know how the high ideals of a free citizenry with a sense of place in a cherished natural environment were variously realized and lost, often to be regained for limited periods of time in the same area or in other parts of the world. We will have to ask how "communal liberty" (to use Benjamin Barber's phrase) has fared with the fortunes of the citizenry and how each interacted with the other, for neither the city's forms of freedom nor the citizen can be isolated from each other without doing violence to the meaning of both. That urbanization eventually separated from citification to take on a life of its own and ravage the city and countryside, ecological as well as agricultural alike will be an abiding theme in all we shall have to explore. This separation begins with the massive institutional, technological, and social changes that eventually dispossessed the citizen of his or her place in the city's decision-making processes. Urbanization, in effect, both presupposes and later promotes the reduction of the citizen to a "taxpayer," "constituent," or part of an "electorate." We shall see that urbanization yields not only a drastic colonization of the countryside but also of the city's and the citizen's very self-identity. Like the modern market, which has invaded every sphere of personal life, we shall find that urbanization has swept before it all the civic as well as agrarian institutions that provided even a modicum of autonomy to the individual. Born of the city, urbanization has been its parent's most effective assailant, not to speak of the agrarian world that it has almost completely undone.

It will be important to see the extent to which presumably non-urban institutions, often remotely tribal in origin and rooted in a more naturalistic society, became integral features of the democratic city in the form of popular assemblies, neighborhood councils, and the town meetings so redolent of an active municipalism and citizenship in our own day. Ironically, the same can be said for feudal, theocratic, and monarchical institutions, with the castle, temple, and palace as their centers. Indeed, we need go no further, if we choose, than the American and French revolutions to find that popular assemblies under different names have been the principal means by which ordinary people fought out the issues of justice and freedom with nobles, monarchs, and centralized nation-states. Each opposing camp has fought with the other for civic and ultimately social sovereignty. The current legalistic image of the city as a "creature" of the state is not an expression of contempt. It is an expression of fear, of careful deliberation in a purposive effort to subdue popular democracy. That the term has been codified into laws, even in self-professed "democracies," expresses a dread of a menacing civic or municipal incubus in every centralized social system, one that has always threatened to dismember centralized power as such and restore the control of society to a public that has been cruelly dispossessed of its very identity.

History is an all-important vehicle in our enterprise, the counterpart of evolution in an ecological approach. The "powers-that-be" live in a compulsive fear of remembrance, a fear of humanity's social memory of past institutions, cultures, and the search for origins. An essential theme of George Orwell's *1984* is the effort by a highly totalitarian state to eliminate the sense of contrast earlier lifeways imposed as a challenge to existing ones. Thinking itself had to be restructured to exclude this challenge by using words—Orwell's famous "Newspeak"—that attenuated their previous wealth of meaning and the disquieting alternatives that the past posed to a fixed, eternalized, and ahistorical "now" or commitment to "nowness." Previously, authority had rested on tradition, often in a highly distorted form; today, it rests on conditioning, with no regard to a troubling past.

The purpose of Orwell's "Newspeak" was to change thinking by "the invention of new words and by stripping such words as remained of unorthodox meanings, and insofar as possible of all meanings whatever." Words such, as "free," Orwell tells us, still existed in "Newspeak" but not "in its old sense of 'politically free' or 'intellectually free,' since political and intellectual freedom no longer existed even as concepts and were therefore of necessity nameless."[1] One could only use the word "free" in a strictly functional, amoral, and technical sense, such as in the statement "This dog is free from lice" or "This field is free from weeds." "Newspeak was designed not to extend but diminish the range of thought" by a process of abbreviation—by rendering concepts more functional than moral in nature and by dememorizing thought (if I may coin a phrase) that was designed to uproot the mind from a sense of continuity and contrast with a challenging past.

From Urbanization to Cities makes no compromises with an emphasis on "nowness" and the cybernetic language of electronic circuitry that is so fashionable today. The pages that follow are thoroughly infused with history and its moral meaning, and with language that is rich in secondary meanings. The reader will find no words like "feedback," "input," "output," and "bottom lines," which are currently used as substitutes for processual and thought-laden terms such as dialogue, origins, "explanations," "judgments," and "conclusions." I have been at pains to emphasize my use of history and traditional language for a very distinct reason. This book deals with cities and citizenship—traditionally, the shared fate that confronts "town and country" in the modern era—and the impact of urbanization upon personality, freedom, and humanity's sensitivity to nature. But it is also a book about morality and ethics. My concern with the way people commune—that is, actively associate with each other, not merely form communities—is an ethical concern of the highest priority in this work. I am concerned with the "social compacts" people form as ethical beings and the institutions they create to embody their ethical goals.

To a great extent, this is the Greek, more precisely, the Athenian, ideal of civicism, citizenship, and politics, an ideal, that has surfaced repeatedly throughout history. I believe this ideal forms a crucial challenge—despite its many limitations for the modern era. I propose to explore not only the ills of urbanization insofar as they have subverted town and country (including the natural environment), alike, but to explore the social and cultural conditions that gave rise to communities, citizens, and a politics whose high regard for personal activism has always made community the richest and most fulfilling expression of our humanity.

Chapter Two

From Tribe to City

Conventional accounts of the city's origins tend today to be stridently technological: they anchor the emergence of the city in the discovery of food cultivation, particularly its highly productive form of animal-powered agriculture. The city, it is assumed, appeared because of the large food surpluses farming folk could provide with the Neolithic technological innovations that marked the cultivation of the land. With this new material plentitude at their disposal, we are told, people began to detach themselves from agricultural pursuits and develop their skills as potters, weavers, metallurgists, carpenters, jewelers, and masons, not to speak of administrators, priests, soldiers, and artists. As agrarian villages increased in size and density, they are said to have reached a "critical mass"—often of undefinable size—that apparently qualified them to be called "cities." Generally, we tend to regard the city itself as a sharp economic breach with the countryside, marked by a typically urban development of crafts, administration, and, to use a grossly denatured word, "politics." This new, largely nonagrarian ensemble of activities produced what we like to call "civilization": a literate world, culturally "enlightened," presumably more rational institutionally and technologically than the agrarian society on which it relied—in short, what the distinguished Marxian archaeologist, V. Gordon Childe, called the "urban revolution."

This conventional image of the city's origins projects a highly modern view, largely economistic and progressivistic, onto the past. It assumes that because we are primarily economic beings whose civic activities are deeply rooted in industrial, commercial, and service occupations, our urban "forebearers" gathered in towns and cities to follow similar pursuits. They, too, we tend to believe, conceived of the city as an economic enterprise, massively committed to nonagrarian tasks. We are prepared to concede that in a more "barbarous" era early city dwellers were also preoccupied with their safety or defense from rival cities or pastoral nomads. Hence, they congregated in great numbers behind defensive palisades and fortified walls. While defensive concerns might account in part for early urban density, they too were a function of economic concerns in modern eyes, just as we, today, assign economic motivations to what we also call defense by nation-states and imperialistic blocs. Thus we assume that our urban "forebearers" were very much like us. They were economic beings who were busily engaged in the pursuit of their material interests within the fixed confines of a structural and territorial entity called the "city."

With equal alacrity we assume that just as they shared our economic lifeways (albeit in a more rudimentary fashion), they also shared our civic attitudes. Although we are likely to concede that their sense of communal loyalties was stronger than ours, we often believe that they judged their cities with a shared viewpoint like ours. However exotic many of their civic institutions seem in the light of our own, we tend to believe that their notion of citizenship was essentially as self-serving and self-interested as our own. They participated in civic affairs to the degree that their material interests were involved, and essentially their interests were no less economic than our own. Unwittingly, we subject their civic-mindedness as well as their "civilization" to the very economistic class analyses we profess to reject in the name of our "higher ideals" and "morals," however much these are honored in the breach.

This view is greatly reinforced by the historical literature at our disposal. Athens, we are reminded, had its *demos*; Rome, its *plebs*; the medieval commune, its *popolo*, just as we have our proletariat and lower middle classes who live in gnawing envy or hatred of their aristocratic and bourgeois elites. We are reminded that such terms as "ancient," "medieval," and "modern" should not impel us to greatly distinguish the unalterable content of human nature, that human beings will seek to satisfy their egoistic impulses despite all their ideological avowals to the contrary. This excursion into an unvarying psyche that lies at the core of human behavior defines the city dweller—particularly the citizen—with the same modern attributes that define our contemporary metropolitan dwellers. Hence, we comfortably sit back before the vast tableau of urban historical development with a sense of self-assurance that our contemporary ills are as ancestral and incorrigibly "human" as our biological attributes and pathologies. They belong to us as assuredly as the human brain, human fingers, and ingrained human psychological traits that our species shares with its ancestors and with its heirs.

If the city provides any evidence of human association and the immutability of human behavior, a serious account of its rise and development in no way supports this simplistic and conventional imagery. It would be difficult to find one all-embracing reason that explains the emergence of a settled human collectivity such as a village, much less a population of thousands that we would expect to find in a city. The earliest cities archaeologists have unearthed do not seem to have been based on advanced forms of food cultivation, notably animal-powered plow agriculture, a point that Jane Jacobs has so ably highlighted in her book *The Economy of Cities*. Although it is doubtful that an "urban revolution" gave rise to an "agricultural revolution," as Jacobs seems to contend, strong evidence exists that such very early cities as Çatalhöyük in Anatolia and Jericho of Biblical fame may have consisted of sizable communities that acquired much, perhaps most, of their food from the hunting of game and the harvesting of undomesticated plants. Such plants as were domesticated seem to have been recent achievements of people who were harvesters of wild wheat varieties rather than experienced food cultivators. Bones of aurochs—the extinct ancestors of modern cattle—as well as those of wild deer, boars, asses, geese, and the skeletal remains of such predators as wolves, foxes, and leopards, suggest a "citified population of hunters and gatherers whose arts of gardening were of recent origin. By conventional standards, this economic tableau does not comfortably explain why a city such as Çatalhöyük, with an estimated population of at least 6,000 people occupying some 30 acres, should have been able to flourish on two nearby sites 9,000 years ago, many millennia before Mesopotamia became the region for Childe's celebrated "urban revolution."

If Çatalhöyük is to be accorded a major place in the origins of the city, the reason for its existence—indeed, for its persistence for centuries—seems primarily to have been religious. Although the city was close to a rich source of obsidian that was almost certainly bartered for a wide variety of nonindigenous foods, it is most conspicuous for the elaborate religious artwork in its houses, paintings on opposing walls that symbolized death in one case and life in another. James Mellaart, who has provided us with the first richly interpretive studies of the city, found no pottery in Çatalhöyük's eastern mound—one of the principal hallmarks of Neolithic culture that, together with plow agriculture and domesticated animals, would normally be associated with a compact city of thousands. The thick walls of Çatalhöyük, its pueblo-like houses, its small rooftop plazas, and its ornate artwork so visible to the public suggest an intensely vivid religious life that is equally suggestive of a communal life. Its tool kit and highly naturalistic artistry suggest an ecologically oriented community of late Paleolithic hunters and gatherers rather than an early Neolithic community of food cultivators. The culture is marked by a very sophisticated stone and bone technics, by markedly collective dwellings adorned with images of animals and shamanlike figures amidst paintings of reindeer, leopards, and bow-carrying hunters.

If we are to judge by the considerable amount of comment Çatalhöyük has elicited, the society may well have been rather egalitarian. Women figure highly in the symbolism of the city's cults and are the most conspicuous figurines that we find among the city's small statuettes. Nor is it clear if hierarchy and warfare were features of the city's social life. Judging from the size of Çatalhöyük's dwellings and the implements found in burial remains, the city was fairly egalitarian despite minor differences that are observable. Evidence from human remains shows a similar quality of life and health for men and women. There are no "obvious signs of violence or deliberate signs of destruction," Mellaart observes for the original city and its nearby successor, both of which appear to have been abandoned after centuries of occupancy.[1] Cases of violent death among the hundreds of skeletons examined on the sites are notable for their rarity.

In fact, it is fair to say that the Çatalhöyük people, who may not have been particularly exceptional, raise the question, Why did "civilization" (more precisely, the drastic change from hunting and gathering to food cultivation and civic social relationships) emerge at all? This question seems shrouded in mystery because the changeover from one cultural form to another has historically been portrayed as more drastic than it actually was. What Çatalhöyük and the earliest food cultivators tell us is that the transition from tribe to village and city was not the predictable result, as archaeological orthodoxy would have it, of technological change or, as some have claimed, of population pressure, war, or other drastic environmental pressures that might have produced hunger on a large scale. In short, the transition from tribe to city was not the necessary result of economistic relationships that our Euro-American minds foist upon prehistory and history; nor was the transition, when it occurred, as deliciously complete and sharply delineated with polarities of a war between town and country as theoretical orthodoxy would have us believe.

Which is not to say that cities congeal out of mere mist. Clearly a city requires a tangible food supply, one that is sufficiently plentiful to support such non-agrarian strata as artisans, administrators, and shamanlike priests to perform their specialties. If the remains of Çatalhöyük suggest anything, however, it is that early cities formed to meet cultural rather than strictly economic or defensive needs. The rich spiritual and domestic lives so evident at Çatalhöyük suggest that spiritual practices were woven into domestic life at every level from sub-floor burials of relatives to the exuberant wall art and do more to explain why this city arose in Anatolia many millennia ago than do economic or military functions. Paleolithic lifeways; richly elaborated by time and environmental changes, may have been more tenacious than we have supposed them to be. They may have been more attractive to tribal peoples, even many self-anointed "civilized" ones, or, at least, more deeply rooted in the long evolution of human culture.

If this conclusion is sound, the urban and agricultural "revolutions" so closely associated in the archaeological literature with the rise of urban culture do not form an elegant fit by modern standards. The rise of cities may have had more to do with shrines, cultic practices, and temples rich in naturalistic symbols than with the "discovery" of cereal cultivation, plows, and domesticated animals. Not that the city gave rise to these agrarian basics, as Jane Jacobs seems to suggest. Apparently, agriculture in a simple form was known to hunters and food gatherers long before villages began to dot the landscape that phased from the Paleolithic into the Neolithic. But the household as a pivot of spiritual and ritual life, later enclosed by a temple, may have been more authentically a harbinger of the city than the plow, and a quasireligious figure such as the shaman or priest may have been an earlier civic leader than the politically astute chief. By the same token, the earliest "citizen" may have assumed his or her civic functions as a member of a congregation rather than, as a "resident" of an urban district. The earliest civic center, in effect, may have been a ceremonial area rather than a marketplace, a center for the worship of natural deities and forces.

If this background accurately portrays the factors that gave rise to the city and the functions of its residents, it highlights many features of early city life that contradict modern economistic biases about urbanism, notably contemporary ones that visualize the city as a business enterprise and its concerns as primarily fiscal or commercial. Indeed, even when the domestic marketplace becomes firmly visible—in Hellenic times, when the Athenian *agora* became a modest center for exchanges of goods as well as intense civic activity—Aristotle was to regard moneymaking as an "unnatural" urge that required public control and self-restraint. What early city dwellers actually exchanged with each other were services. More precisely, men and women cojointly contributed their share to the common fund of material goods. It was this shared pool of the means of life that constituted the "economic life" of tribal communities and early cities. People contributed, in effect, their skills in growing food, in working wool and flax, in crafting metals and stone, and, by no means of least importance, their artistic and decorative talents in bejeweling and designing artifacts for deities, priests, and members of their own community.

"Postulates" of "self-sufficiency" and the distinction between "natural and unnatural trade" are not strictly archaic. We find them most self-consciously and philosophically stated in Greece. But prior to the full flowering of the Hellenic *polis*, material life in cities was deeply embedded in the blood ties, religious obligations, mutual loyalties, magical techniques, and the intensely naturalistic sensibilities of the tribal world. Guided more by custom than rationally voiced strictures, these form the psychological setting and institutional carryovers for the more rationalistic civilization we later find in the Greek archipelago. This unconscious tribal and mutualistic substrate of obligation and association was, in fact, more forceful as a guide for human behavior than its formulation into a sophisticated civic and political philosophy or social theory.

In the so-called "primitive" or archaic worlds, more than food entered the "common pool." Possibly, most things short of one's closest possessions, including aspects of one's very identity, had a highly collective aura that destined them to be shared or to be used communally. Initially, if the city had a pronounced function at all, it was a religious one. The emphasis that some researchers placed on political and centralized governmental forms as institutions for efficiently redistributing produce from ecologically different areas may be overstated. That great imperial systems, such as those of the Incas, Aztecs, Babylonians, Egyptians, Persians, and Chinese, were centers for collecting and redistributing a great variety of goods from highly diverse and distant ecological regions can hardly be faulted as a reality. Indeed, by imperial times many ancient cities were as conspicuous for their warehouses as they were for their temples and palaces.

But such an emphasis on material function, like the highly deterministic strategies to explain the existence of every community within our purview, betrays a very modern bias. It reveals a proclivity, almost an unthinking compulsion, to assign a "material role" to every phenomenon, particularly every institution and form of association, that exists in the past as well as the present. Worse, it crudely downgrades the richly associative role played by material things such as gifts, which serve to foster a much deeper human attribute: the need to be grounded in community, to enjoy shared sensibilities that are spiritually supportive and without which authentic individuality is chimerical. Human personality, which is nurtured by parental care, kinship ties, friendship, and the assurance or security provided by personal support systems, becomes a material thing—a manipulatable object among many other objects and commodities—precisely when its immaterial support systems are subverted and its traits reified. Quite an opposite case can be made for the belief, so widely promoted by contemporary "cultural materialists," that chiefdoms, monarchies, bureaucracies, armies, clerical hierarchies, and the unlimited investment they required were distributive agencies and processes that served humanity's "needs," mythic or real. It could be more validly shown that they reflected a mania for domination that created mythic "needs" and systems of control on a scale so harmful to the communities they were pledged to service that they and their legacy of waste, destruction, and cruelty now threaten the very existence of society and its natural fundament. Indeed the domination of nature was to have its roots in the domination of human by human such that a credo of domination was to embrace the planet.

But an important caveat must be voiced here when we speak of a mania for domination that can so facilely be used to color our image of the early, essentially temple, cities at the dawn of civic life. By no means is it clear that the sacerdotal hierarchies that emerged in these cities from Erech in Mesopotamia to Teotihuacan in Mexico immediately led to a hierarchical restructuring of the fairly egalitarian tribal or village peoples on whom they depended—peoples who built their monuments, often massive in size; who created their plazas, dwellings, and altars; who filled their temple storehouses, erected their walls, and shaped the sculptures that dazzle the modern eye. That such vast efforts with their enormous mobilizations of labor could have been made for purposes that seem so "profitless" and "useless" by present-day urban standards without coercion of the most authoritarian kind seems unthinkable at first glance—so much, in fact, that all early temples and mortuaries are normally regarded as the work of savagely coercive rulers and brutal tyrants.

But if this imagery is certainly true well into history, we have no reason to believe that it reflects the social relationships that gave us our earliest cities and their cultic structures at the dawn of history. I have described elsewhere, in great detail, how an egalitarian society may have slowly phased into an increasingly hierarchical one—initially theocratic, ultimately feudal, and finally monarchical.[2] Here, I would like to emphasize that the earliest cities were largely ideological creations of highly complex, strongly affiliated, and intensely mutualistic communities of kin groups, ecological in outlook and essentially egalitarian and nondomineering in character. We do not know if a sizable city like Çatalhöyük, which dates back 9,000 years, or a hugely monumental city like Teotihuacan, which was slowly erected around 300 BCE and ceased to be occupied around 800 CE, were originally constructed by forcibly "mobilizing" large numbers of "oppressed" villagers in surrounding communities or, surprising as it may seem, whether they were voluntary enterprises undertaken by devout "parishioners" who viewed their civic responsibilities as a sort of "calling." We assume that a coercive strategy was followed by oppressive elites at the inception of city life because we read our literary accounts of Mesopotamian and Egyptian forced labor back into city lifeways in a misty preliterate era. It is easy to overlook the fact that any literary tradition of urban life, even the very early Gilgamesh epic from 2100 BCE which dates back to the beginnings of Mesopotamian city life, is already evidence of a technically advanced, often coercive, society. In contrast, Çatalhöyük, Jericho, Erech, Teotihuacan,

Monte Alban, and Tikal, to cite cities far-removed from modern urban development in the cultured areas of the Near East and the Americas, are mysteries to us that we try to dispel with our own motivations and social interests. Yet their archaeological remains are rich with the evidence of ideological aspirations and human relationships that were fundamentally different from our own secular ones—cultic and communal sensibilities that viewed "the city" (if such a word can be used indiscriminately to encompass all sizable human settlements) as a monument to lifeways and sentiments that were basically different from our own. Possibly, the ordinary people who reared the monuments in the earliest of these cities were closer in their outlook to the medieval artisans who often willingly gave of their time and skills to slowly erect over generations the great cathedrals of Europe. In either case, they speak to a dedication that stands in marked contrast to the mentality of the engineers and construction crews who are seeding the modern world with high-rise residential and office structures in urban areas of every continent.

We are confronted, too, with the differences between early and modern conceptions of "citizenship." How did ancestral "citizens" of the first cities view themselves? In what sense were they different from us—or similar to us? By asking these questions, we encounter a problem that lacks the degree of completeness and fixity that we find in a physical structure such as the remains of a temple or palace. Citizenship, as we shall see, is a process, not a reality that is reducible to a concise, single-line definition. It does not leap into being from a vacuum, surrounded by streets for personal display, buildings into which the sovereign individual can retreat, and dense populations that foster personal intercourse. Çatalhöyük, in fact, had no streets at all, to cite an intriguing feature of the city. A pueblolike city, it had small squares but no open byways One moved from one part of Çatalhöyük to another over rooftops, ascending or descending ladders, entering into the recesses of dwellings and crossing small squares.

This is a significant personal fact, not merely a structural eccentricity. Çatalhöyük was intensely peopled—more like a tribe than what we, today, would call a city. Yet even by modern standards of urbanism such as size and density, it was a town, not a village. From a historical perspective, given time and place, it was even an immense city if archaeological calculations of its populations are remotely correct. What does this tableau mean? In what sense were its men and women "urban"? We can surmise that they were not mere "residents" of the city or the ancestral members of modern-day "constituencies." They were not burdened by the anonymity and awesome sense of personal isolation that is the most characteristic trait of the modern urban dweller. In some sense, they were "protocitizens" of a highly articulated and richly textured community in which a high sense of collectivity, nourished by such organic facts as kinship ties and a sexual division of labor, was integrated with the civic facts of politically defined rights and duties. They were communities in transition between the biological realities of the tribal world, rooted in blood ties, gender, and age groups, and the political realities of the urban world, rooted in residential propinquity, vocational mobility, and legal prerogatives. Early cities probably did not contain citizens in the sense of self-empowered individuals ethically united by ideals of civic virtue, rational in their social policies, and completely free to participate through discourse and practice in the management of their cities—in short, those attributes that Greek social thinkers were to call *phronesis*, the practical reason involved in creating and managing a community. All the evidence we have of these protocitizens suggests that their power for social action was largely controlled by obligations to kin groups and theocrats, their ideals guided more by faith than reason, their sense of virtue more pragmatic than ethical, and their social institutions more biologically derivative than cultural in character.

Yet these seemingly uncivic features provided a crucially important matrix for what was to grow into the highly sophisticated classical notion of the "citizen." If social empowerment seemed to derive more from such group attributes as the family or the clan than from personal attributes, the individual living in these cities enjoyed a real sense of power as such, not a body of conferred rights that were more formal and juridical than substantive. Tribal societies are known that exhibit a high degree of respect for individual uniqueness and free will, however much custom and public opinion seem to place limits on personal behavior. One cannot simultaneously deny the existence of primitive individuality while acknowledging the existence of a fairly spontaneous ego and considerable self-assertion among, say, certain hunting and gathering communities such as the pygmies of the Ituri Forest who include outrageously boastful men and extremely shrewd women. By their marked presence, such personal traits as boastfulness and shrewdness, indeed humor, gaeity, and reflectiveness, frankly contradict the conventional image of preliterate peoples as divested of ego and personality, the modern claim to individuality. The contemporary neurotic notion of personality may not have been as common among so-called primitives as it is today in the metropolitan areas of Europe and America, although by no means is it absent. But individuality certainly exists among remaining preliterate communities, albeit in a different form and faced by more overt constraints than our own, notably where the rules of the game are fairly explicit and mutualistic in contrast to our modern, highly engineered society, which bombastically celebrates its formal "freedoms" and tries to ignore its lack of social concern.

Surprisingly, citizenship and the political forms that foster it would be difficult to explain without looking precisely at those primal organic institutions—particularly clan-like relationships, popular assemblies or the various councils that spin off from them, and the egalitarian outlook, in sum, the tribalism—that the city, in the conventional wisdom of the recent past, was supposedly designed to overcome. Rousseau was to pithily observe that "houses make a town but citizens make a city."[3] Yet he would have been no less surprised than all the major social theorists of his day to learn that the most important attributes of citizenship derive more directly from the tribal world than the village world, from rude shelters rather than houses. Houses may make towns, even huge urban belts that subvert the conditions for an active, participatory body politic, but simple huts and tents—the "primitive" dwellings of hunting, gathering, and pastoral peoples—provided the homes for institutions and social types that often embodied the civic ideals we associate with citizenship.

Notions of human scale—of communities that are modest in size and comprehensible politically and logistically to their residents— are distinctly tribalistic in character and origin. They are formed from the idiom of a civic mentality that is rooted in familial loyalties and extended kinship relationships. Not surprisingly, many early cities, insofar as they leave any written account that is open to us, are "founded" entities. Their "ancestry" originates in a shared deity or a delineative progenitor with some kind of personal embodiment. The Sons of Aeneas, the Trojan hero who "founded" Rome in the city's urban mythology, are no different in principle than the Children of Israel and the Biblical Jacob who was the patriarchal ancestor of the Hebrew tribes. The word "citizen," in fact, appears sporadically or late in the history of cities. Quite commonly, the "citizens" of a community denoted themselves as "brothers," a term in widespread use throughout the Middle Ages and early Renaissance. Men and women in the towns and cities of the past visualized their relationships in terms of familial connections. As "strangers" began to form the majority of urban dwellers in late classical and medieval cities, this familial imagery with its emphasis on smallness of scale, accessibility of person, and close-knit support systems of the kind we associate with "humanly scaled" communities became the outlook and prerogative of urban elites, notably the municipal aristocrats and nobles who staked out a real or legendary claim to the city's "founder." Ultimately, the newer dwellers of the city, too, formed their own "brotherhoods" in which ties, rights, and duties were solemnized by blood oaths and kinshiplike rituals. In time, the word "brother" became an ecumenical form of civic address and affiliation, spanning class ties and interests.

The civic institutions that we most commonly associate with a "participatory democracy" often reach back in almost unbroken continuity to tribal assembles. It is fairly certain that face-to-face assemblies of the people in ancient cities were often the keystone of the civic institutional arch and extensions of tribal assemblages with their remote origins in primitive egalitarian relationships. So, too, the notions of consensus where it existed, prolonged discourse with its goal of arriving at a commonality of views, and a sense of agreement were pronounced features of such a radical political democracy. By contrast, councils, the representative organs of assemblies, patently originate in the smaller assemblies of elders, chiefs, and warriors. As contractions of the popular assembly—the "voice of the people" as distinguished from the people themselves—they are ubiquitous features of a tribalistic view of the reduced assembly projected into republican concepts of self-governance.

Yet here, too, I must add a caveat that initially seems to break the continuity of the development from tribe to city. The city may continue the traditions of the tribe institutionally into a rich skein of participatory, indeed ecological relationships, but it also constitutes its antithesis biologically. The city is the perpetuator of the kin group insofar as the latter is a parochial expression of blood ties that exclude the stranger—the outsider—who cannot claim a common ancestry with the brothers and sisters. Tribalism is equatable with familial exclusivity, not only familial solidarity. The distinction between exclusivity and solidarity is crucial here. Whether fictive or real, tribalism has its universal solvent in blood, the medium that accords equality or *isonomia* (to use the Hellenic term) to its members. Theoretically, one does not join a tribe; one is born into it or, at most, adopted by it because of the services it requires. Adoption involves elaborate rituals that transform the outsider into an insider—an "inorganic" being, to use Marx's formulation, into an "organic" part of the community.

The city breaks this biological spell, with its ecological aura, however fictive it may be in reality. It exorcises the blood oath from the family with its parochial myths and its chauvinistic exclusivity, while retaining or reworking its concept of socialization. Paternity is given a high place in the civic firmament of values, but the fathers are slowly divested of their absolute powers over the sons—sons who are needed by the city to perform administrative and military services. The stranger is denied formal legal status in the city's system of governance but is admitted into the magic circle of civic protection and solicitude, especially if bearing skills, wealth, and material resources.

But the biological aura remains as legend by hypostasizing the city's founders, the ethnic continuity of its citizens, the status of its rulers, although reality enhances the hidden powers of the newer residents who can claim no tie to the civic progenitors. Hence, the city creates a special kind of social space. A projection of familial and tribal forms, the city subverts the authentic substance of the parochialism that flavors kinship ties with a mysterious inwardness and rescues tribal institutions as ecumenical forms of civic administration. The Greek democrat, Kleisthenes, is almost a symbol of these shrewd maneuvers, maneuvers by no means unique to the Athenian *polis*. To break the hold of family ties that obstructed the power of civic institutions, the citizenry was organized into territorial "wards," but each "ward' was felicitously called a "tribe." The municipal space of Athens, in effect, was expanded to create a largely civic citizenry, unencumbered by the mindless tribal obligations and blood oaths that impeded the rights of the stranger but in a form that wore the symbols and enjoyed the prestige of tribal tradition. Indeed, one of the great tasks of ecological thinking will be to develop an ecological civicism that restores the organic bonds of community without reverting to the archaic blood-tie at one extreme or the totalitarian "folk philosophy" of fascism at the other.

Chapter Three
The Creation of Politics

Politics has acquired a fairly odious reputation among the great majority of people today. The word seems to denote techniques for the unsavory end of exercising power over human beings. We "play politics" not only on an international, national, and local scale; we do so in domestic relations, in schools and places of learning, in ordinary jobs or extraordinary careers. Politics, in effect, is seen to have invaded the most private recesses of our lives. At worst, it is viewed as oppressive, manipulative, cunningly seductive, and basically degrading. Few words more readily evoke a contemptuous sneer than the term "politician." Conceived instrumentally as a means to control people, politics is regarded as inherently corruptive of both its user, the politician, and the public on whom it is used. The ideal of a political life earns few acolytes among people with any degree of moral probity. Traditional conservatives and anarchists alike identify it with the state and preach a message of the attenuation or outright abolition of "political power." Liberals and socialists rarely celebrate politics as a desideratum, but they wed it so closely to the state—a necessity in modern liberal and socialist theory—that its practice is seen as unavoidable in a highly imperfect world.

Modern social ideologies tend to blend politics with the state almost unthinkingly—and often throw society into the brew for good measure. The confusion on this score is massive. Just as there are many people who, by virtue of the all-pervasive role the state plays in their private lives, draw no distinction between "government" and "society," so an incalculable number are incapable of distinguishing between the state and "politics." These attitudes—and they are often little more—are justified today by ordinary experience, as we shall see. That the social, political, and "statified" are not synonymous, indeed, that they are arenas that can be broadly demarcated with histories and identities of their own, is so far removed from the public's mind that to distinguish them seems paradoxical.[1] The equatability of political with state activities is taken as a given. The penetration of the two, normally conceived as one common phenomenon, into private affairs still encounters considerable resistance, although in the form of a psychological tension that finds expression in existential resistance such as petty violations of the law rather than ideological clarity.

In recent years, however, serious attempts have been made to probe the distinction between "society" and "politics," which has traditional roots in theoretical distinctions between society and the state. Anarchists have been saying for years what everyone either knows or feels: the state is not the same kind of phenomenon as the family, workplace, fraternal and sororal groups, religious congregations, unions and professional societies, in short, the "private" world that individuals create or inherit to meet their personal and spiritual needs. This personal world can be designated as "social," however much "government" penetrates, regulates, or, in totalitarian states, absorbs its forms. It is a world that has deep roots in what Marx and later Hannah Arendt were to designate as the "realm of necessity," a world where the individual satisfies the conditions for his or her personal survival. Here, biology provides the soil for a system of self-maintenance such that people have a culturally conditioned but systematic way of reproducing their kind, feeding and clothing themselves, satisfying their needs for shelter and the support systems for resisting an inclement, presumably "cruel, mute, and blind natural world."

To have sharpened the distinction between the social and the state is one of the major contributions of traditional anarchist theory. That there could be a political arena independent of the state and the social, however, was to elude most radical social thinkers, even so intense a political thinker as Marx who allowed for "democratic" states that could "evolve" toward socialism and "Bonapartist" states that stood above and balanced off conflicting class interests. The reformist wing of Marxian socialism was by no means alone when it envisioned a pliable state that could be used for socialist interests. Its own "founding fathers" were no less riddled with uncertainties about the nature of the state than the reformists are today.

The emergence of the political realm as unique has a complex background in the history of ideas. Politics as a phenomenon distinguishable from the state and from social life initially appears in the extant writings of Aristotle, perhaps the most Hellenic of the Greek social theorists and philosophers. With Aristotle we are still dealing in terms of human association on the level of the city, or to be more precise, the *polis*, which is commonly mistranslated as the "city-state."[2]

By the middle of the fifth century B.C., when the Athenian democracy was approaching its high point of development, the concept of a state—of a professionalized bureaucratic apparatus for social control—was notable for its absence. Attic Greek contains no word for state. The term is Latin in origin, and its etymological roots are highly ambiguous. It more properly denotes a person's condition in life—his or her status or way of life and "standing"—than a commonwealth or a state in the modern sense of the term. Not until the early sixteenth century, when we witness the rise of authentic nation-states and highly centralized monarchies, does the word come to mean a professional civil authority with the power to govern a "body politic."

There is a very real sense in which the evolution of the word broadly reflects the evolution of the state itself. Not that state powers were rarities in the ancient and medieval worlds. Eric Vogelin's "cosmological" empires—Mesopotamian, Egyptian, Persian—and the more "ecumenical" Roman empire were certainly states in the sense that they controlled vast resources, dominated millions of people,

and were structured along highly professional, rationalized, and bureaucratic lines. State institutions emerged very early in human history, although in varying degrees of development and stability, often
in "bits and pieces," as it were, with highly tribalistic features. The
Pharaonic state in the Nile valley already reached back thousands
of years, perhaps beyond the unification of northern and southern
Egypt, long before the Roman empire had begun its long decline.

Athens had a "state" in a very limited and piecemeal sense. Despite its governmental system for dealing with a sizable slave population, the "state" as we know it in modern times could hardly be
said to exist among the Greeks, unless we are so reductionist as to
view any system of authority and rule as statist. Such a view would
grossly simplify the actual conditions under which humanity lived
in the "civilized" world.[3] Until recent times, professional systems of
governance and violence coexisted with richly articulated community forms at the base of society—city neighborhoods in the world's
few large urban areas, self-contained towns and villages, a network
of extended kinship ties, a great variety of vocational, mutual aid,
and fraternal groups—which were largely beyond the reach of centralized state authorities. In fact, these distinctly social formations
were necessary to the maintenance of the state. They were sources of
its revenues, its military personnel, and, in many cases, the source of
many labor services for a great variety of public and religious tasks.

The Athenian democracy, if anything, was the opposite of a professionalized system of governance organized strictly for social
control. If we choose to translate the word for the Athenian *polis* as
"state," which is done with appalling promiscuity, we would have to
assume that the notion of a state is consistent with a body politic of
some forty thousand male citizens, admittedly an elite when placed
against a still larger population of adult males possibly three times
that number who were slaves and disenfranchised resident aliens.
Yet the citizens of Athens could hardly be called a "class" in any
meaningful sense of the term. Indeed, within this body politic, we
encounter economic distinctions that run the entire gamut of material resources from the wealthy to the impoverished.

We would also have to assume that the notion of a state is

consistent with a consciously amateur system of governance, based on almost weekly popular assemblies, a judicial system structured around huge juries that represent the assemblies on an attenuated scale, the selection and rotation of civic officials by sortition, that is, the use of the lot, and the absence of any political professionalism or bureaucratism, including military forces that are authentic militias of armed citizens rather than professional soldiers.

These few remarks do not do full justice to the Athenian system of democracy, notably its high level of consciousness, civicism, commitment, and aesthetics. They must suffice for the present to understand how politics was created on the rare occasions when it appeared in history with a reasonable degree of authenticity, and how it was conceived by its most renowned theorist, Aristotle. It is to Aristotle that we must turn for the earliest known distinctions between the social and political realms, the household and the public arena. The state was not alien to Aristotle; it existed as monarchy and tyranny, as everyone could see among the states that surrounded the Athens of Pericles's day. And they were as easily confused with the *polis* by Aristotle and his students, just as we, today, confuse *politikos*, the administration of the *polis*'s affairs, with the state. What was central to Aristotle's thought, however, was the *polis*, not that ambiguous phenomenon, the state. What is central to modern thought, by contrast, is the state not politics or, strictly speaking, the affairs of administering an entity that could pass for a *polis*. Hence, although politics and the state intermesh in both cases, they do so in very different ways and from remarkably different perspectives.

This distinction should not be reduced to a simple difference of vantage points toward a shared phenomenon. Aristotle and the modern social theorists are, in fact, looking at two very dissimilar worlds. The difference becomes very evident in Book Seven of Aristotle's *Politics* where the proper size of the "ideal" *polis* is discussed. Both Aristotle and most modern social theorists would agree that a community that is very small risks the possibility, if it relies primarily on its own resources, of inadequately furnishing its inhabitants with the means of existence, much less providing for the "good life," however one chooses to interpret this highly problematical phrase. Almost presciently, Aristotle derides the view that a community's greatness is to be judged by its demographic and territorial size. Important as numbers and resources may be, a *polis* with too large a population and area cannot have a "good legal government.... Law [*nomos*] is a form of order, and good law must necessarily mean good order; but an excessively large number cannot participate in order: to give it order would surely be a task for divine power, which holds the universe together. Hence the *polis* also must necessarily be the most beautiful with whose magnitude it combines the above-mentioned limiting principle; for certainly beauty is usually found in number and magnitude, but there is a due magnitude for a *polis* as there is for all things—animals, plants, tools," otherwise it will lose "its true nature" as well as workability. Summing up this remarkable body of notions, Aristotle concludes "that the best limiting principle for a *polis* is the largest expansion of the population with a view to self-sufficiency that can be taken in at one view."[4]

Probably no modern social theorist would reason out this case for "beauty" *qua* "magnitude" and "limit" *qua* "true nature" as the essence of a community. The case for human scale has been argued by heterodox urbanologists on logistical, democratic, and aesthetic grounds; but rarely, if ever, has it been argued on ethical, indeed metaphysical, grounds. A community "that can be taken in at one view"—that is, decentralized, comprehensible, and attractive—which Aristotle and his modern counterparts regard as a desideratum, is rooted in many similar but also many different premises. Aristotle speaks to us from an age that found in magnitude and harmony the essence and, in the "true nature" of a human community, the limit to a *polis*. An ethical pragmatism pervades Aristotle's remarks that qualitatively differs from the instrumental pragmatism of the modern urbanologist, however much the two share a common practical view of human consociation.

Politics, in turn, is also inseparable in Aristotle's mind from its ethical context. Men are "animals," a fact that greets us early on in the *Politics*, but they are animals of a very special kind. It is man's destiny or *telos*, if he is to fulfill his "true nature," to live in the *polis*. A *polis*, however, is more than a community or *koinonia*. It is a *koinonia* that has reached the ideal form of a shared commonality of purpose among men whose self-realization is the "good life."

The "good life," in turn, includes a degree of material self-sufficiency that goes beyond mere survival. But it does not consist in an appetite for goods, with all its attendant excesses, that clouds man's ethical and intellectual clarity. Man transcends his animality insofar as he has reason and speech, or *logos*, which combines both attributes in the ability to symbolize verbally and generalize logically. But these abilities do not guarantee that man has reached or even approximates the fulfillment of his potentialities. Institutions must exist that constitute the means for achieving human self-fulfillment; a body of ethics must exist that gives the required institutions substance as well as form; a wealth of social activities must be cultivated in the civic center or *agora* of the *polis*, the gymnasium, and in the theater as well as the popular assembly and courts to nourish interactions and discourse; a mode of character development and education, both of which are combined in the Greek word *paideia*, must be at work to enrich the interactions among men and thereby foster the growth of ethical and intellectual insight.

Underlying these various "means" is Aristotle's emphasis on human solidarity or *philia*, which includes friendship (the common English translation for the Greek term) but which is a word more far-reaching in its connotation of civic commonality. The intimacies of friendship may be reserved for a limited few, but *philia* implies an expansive degree of sociality that is a civic attribute of the *polis* and the political life involved in its administration. Man is "by his nature" a political animal or *zoon politiken*, which is to say that he is destined not only to live in a community but also to communize. In criticizing Lykophron the Sophist, who contends that the *polis* is a "mere alliance" among men to prevent them from inflicting harm on each other (Hobbes's later view of the "social contract") and promote the exchange of goods to satisfy their individual needs, Aristotle argues that the *polis* is an end in itself, the realization of man's need for consociation apart from its material benefits. "If men formed the community and came together for the sake of wealth," he declared, "their share in the *polis* [would be] proportionate to their share in the property, so that the argument of the champions of oligarchy would appear to be valid ... but if on the other hand the *polis* was formed not for the sake of life only but rather for the good life and if its object is not military alliance for defence against injury by anybody, [and] it does not exist for the sake of trade and of business relations," the *polis* would be more than a community and its citizens would "take civic virtue and vice into their purview." Indeed, lacking a concern for "civic virtue and vice," men would form communities no different from those of animals or slaves, who are simply concerned with survival. Communities united by mere economic and military alliances—and, here, Aristotle has the Etruscans and Carthaginians in mind—would be no different from the associations other people establish who, for all "agreements about imports and covenants [to abstain] from dishonesty and treaties ... for mutual defence," have no "officials in common" and take no "concern as to the moral character of the other."[5]

By contrast, the household is the sphere of mere survival, the place to which our *zoon politiken* repairs to satisfy his biological need for food, clothing, shelter—in sum, the "realm of necessity" to cite Marx's commonly used phrase. It is the domain of the man's wife, children, kin, and slaves where an apolitical "kingship" (patriarchy) prevails. Here, the man's relationship between his own person and the members of his domestic group is determined not by *logos* but by need, and the social tie is strictly one of "ruler" and "ruled." "The family," we are bluntly told by Aristotle, "is the association established by nature for the supply of men's everyday wants."[6]

But men aspire for more and accordingly group their families together to form villages. To the extent that villages are transformed by man's potential for the good life into ethical and cultural communities, the *polis* begins to appear. The family still exists to satisfy man's animal wants. Hence the two worlds of the social and political emerge, the latter from the former. Aristotle's approach to the rise of the *polis* is emphatically developmental and, in this way, resembles Plato's account of the rise of the "ideal *polis*" in *The Republic*.[7] The *polis* is the culmination of a political whole from the growth of a social and biological part, a realm of the latent and the possible. Family and village do not disappear in Aristotle's treatment of the subject, but they are encompassed by the fuller and more complete domain of the *polis*.

The distinction between the social and political in Aristotle's thinking is strikingly processual: the difference is explained by the growth and development of the social into the political, not by their polarization and mere succession. The state has not yet emerged in a form that gives it a uniqueness apart from the other two domains. "Rule" properly belongs to the family. Where it does appear in the *koinonia*, it is simply a brute extension of the patriarchal family to the civic world (*monarchia*) or it takes the form of a despotism ruled by a tyrant (*tyrannos*). And Aristotle views monarchies and tyrannies as warped or unfinished forms of civic administration that are unbecoming to a *polis*, although rule by a monarch with its traditional constraints on the ruler is to be preferred to rule by a tyrant, which is the arbitrary supremacy of one man.

Nor does he prefer the rule of the few over the many—an aristocracy at best and an oligarchy at worst. By the same token, democracy that Aristotle understands to be the rule of the many over the few—specifically, a condition where the "poor" rule over the "wealthy"—is by no means desirable, although he does not seem to find it as abhorrent as arbitrary one-man rule.

The best-ordered *polis* is structured around a system of governance where the most ethically and materially meritorious stratum of the population manages the *polis*'s affairs in the interests of all. This "polity" or "meritocracy," as it has been called, is an ethical union that simultaneously yields the "good life" in a moral and material sense. Politics consists of the practical reason (*phronesis*) and action (*praxis*) that enters into such a felicitous *koinonia*.

Athenian politics was nothing if it was not vital, indeed voluble, and popular in every sense of the term. Within a span of some three centuries, the Athenian people and their renegade aristocratic surrogates such as Solon, Kleisthenes, and Perikles were to dismember the traditional feudal system of Homeric times, wage a steady war against privilege within the citizen body, and turn the popular assembly from a lifeless, rarely convened mass meeting into a vital on-going forum for making major decisions, thereby opening public life to every Athenian adult male. Power ceased to be the prerogative of a small, well-born stratum of the population. It became a citizen activity. Athens's historic calendar is marked by seething upsurges of the people, startling fluctuations between aristocratic rule, tyranny, limited popular government, until, by the latter half of the fifth century B.C., Athenian political life stabilized around a face-to-face democracy of the most radical kind. We may assume that similar developments occurred in many Hellenic *poleis* that were to ally themselves with Athens for internal political reasons as well as mutual defense.

In any case, in recorded history we have no structure—and I mean simply structure, in view of Athens's many social failings—comparable to the Athenian democracy. Popular assemblies such as the New England town meeting and the Parisian revolutionary sections of 1793–94 were to appear elsewhere over time. The Swiss Confederation is one of the few among many aborted or incomplete examples where popular control formed the underpinnings of an on-going political system. Athens, however, is unique historically in that the *polis* fostered a degree of citizen participation not only in the decision-making activities of the assembly but in the everyday politics of the *agora* that impelled its admirers over the ages to regard it with uncritical adulation as evidence of a "pure" democracy—and its opponents as evidence of a horrendous "mobocracy."

If politics is taken to be a form of popular activity in administering public life that, strictly speaking, is neither a state, conceived as a highly professionalized system of governance, nor a "society," conceived as forms of personal association for promoting survival and well-being, the Athenians' could be said to have literally created politics. It was a parochial politics by modern "global" definitions of the term: civic rather than regional in scale, limited to a minority of the population, Hellenic in its purview of the "civilized" world, contemptuous of slaves, women, the "barbarians" beyond the confines of the Greek *ethnos*, and resident aliens who performed much of skilled work and engaged in most of its trading activities.

Qualitatively, however, Athens made up in depth what it lost in scope. It may well be, as the Jewish Zealots were to believe, that a special insight of a spiritual or moral nature is the privilege not of great empires but of small communities on the margins of the great classical ecumenes. Both peoples, the Hellenes and the Israelites, provide visible evidence of the truth that may be hidden in this seemingly self-serving conviction. Yet a "creation" had certainly emerged that opened a new dispensation in human affairs. A new realm of life had appeared, the political realm, which was to acquire many different meanings but whose origins in classical Greece still keep faith with the pristine values and practices that impart meaning to those ill-used words, the "public sphere."

The Roman Republic, which looms high in its impact on the Euro-American political tradition, stands in marked contrast to the Athenian notion of a public sphere. Polybius, the Greek chronicler of Rome's rise to world hegemony, offers us the classical theory of republican government, a theory that was to deeply affect the thinking of American and French constitutionalists in the eighteenth century.

According to Polybius, the virtues of monarchy were embodied in the consuls, the two chief magistrates of the Republic. The Senate provided the Republic with the advantages of aristocracy, with its gradations in descending order of consular, praetorian, aedilitian, tribunitial, and quaestorial ranks. The image of the Roman Senate freely debating public issues is essentially a myth: no senator could voice his views on an issue unless the presiding consul solicited his opinion, and these requests were directed in a strict hierarchical sequence that often left little time for oratory by the body's lower orders. Finally, democracy was represented by several assemblies of the people. These assemblies (some four have been identified) reflected, in their variety and ascendancy, the fortunes of the plebeians and other lowly orders in their conflicts with the ruling patriciate. Whatever their origins, Roman popular assemblies elected all the magistrates of Rome, some of whom, such as the tribunes and praetors, had enormous control over other branches of the government during their heyday.

Assemblies could be used by ambitious politicians to bypass the Senate and enact laws that the ruling oligarchies opposed. Hence, two political strategies existed during the more fervent periods of the republican era: an oligarchical one and a popular one. The *comitia centuriata* formed the principal popular law-making body of republican times. A complex mix of weighted voting groups sorted out according to military status and, later, according to classes based on property and age, the assembly was highly structured along hierarchical lines. It elected Rome's consuls, praetors, and censors—each successively forming the most important or prestigious magistrates of the republic. The elective functions of the *comitia centuriata* were to be slowly supplanted by the Roman tribal assembly, the *comitia tributa*, largely based on territorial divisions in which 35 tribes were classified into 31 rural tribes and 4 urban ones. Coexisting with both of these ranked assemblies, the *plebs* had their own exclusive *concilium plebis* from which all patricians were excluded. The *concilium* chose its own tribunes and aediles, the latter constituting officials who administered public works and police and took charge of the grain supply and games that gave the city such ill-famed distinction.

We owe the word "plebiscite" (*plebiscitia*) to the right of the plebeian tribunes to submit laws for the approval of the *concilium*. Either the *Campus Martius* along the westward bend of the Tiber or the Forum constituted the principal meeting places for these assemblies, and discussion, if it occurred at all, was minimal. Laws, edicts, declarations of war were presented to the people by officials such as consuls, praetors, and tribunes; elections and confirmations were voted upon methodically. We cannot say with certainty that Roman popular assemblies were simply mute. Before the assemblies divided into their specific units, the highly structured *comitia* was preceded by a loosely organized *contio* where discussion may have been possible. "The *contio* may be a survival of an early form of assembly," opined the late Lilly Ross Taylor, "like that of the Homeric warriors or the Spartans of later times, in which men expressed their opinion by shouting."[8] If so, the *contio* does not bring the Roman assembly form any closer to the highly talkative Athenian assembly. Indeed, Greek observers of Roman procedures found the difference between the *contio* and *comitia* confusing. "With no such distinction in Greek lands between meetings for speaking and those for voting," Ross notes, "Greek writers on Roman institutions have difficulty with the word *contio*."[9] Polybius and Dio Cassius were to call the assemblies by one word, *demos*, or, simply, the people. Appian, an Alexandrian Greek, and Plutarch, who was born in Boeotia, seem to have been removed sufficiently from the traditional *polis* to designate the Roman assembly form as an *ekklesia*, the Greek word for the popular assembly that, as Ross emphasizes, "combined speaking and voting."[10] If the *contio* was anything like the Homeric and Spartan assemblies, it did more listening than speaking, voicing its approval by acclamation and shouts rather than by oratory.

Polybius's association of the Roman assemblies with democracy is specious. The republic had no democratic component in the Hellenic sense of the term, and speech, while relatively free, was more an affair of delivering elitist rhetoric to manipulable audiences in the Forum than the verbal interchange of political equals. A face-to-face relationship between active citizens for the purpose of arriving at a consensus is alien to republican systems of government. A democracy is participatory; a republic, representative. The first involves the exercise of power directly by the people; the second, its delegation to selected surrogates, who then *reconstitute* the political realm that initially existed at the base of the *koinonia* into a distinctly separate and usually professional power at its summit. Republics are beyond the immediate reach of popular control; democracies are not even confronted by the issue of the displacement of power.

Rousseau, with barely concealed irony for the French *philosophes* who were so endeared to English constitutionalism, was to draw these distinctions sharply. "Sovereignty, for the same reason as it makes it inalienable, cannot be represented. It lies essentially in the general will, and will does not admit of representation: it is either the same, or other; there is no intermediate possibility. The deputies of the people, therefore, are not and cannot be its representatives: they are merely its stewards, and can carry through no definitive acts. Every law the people has not ratified in person is null and void—is, in fact, not a law. The people of England regards itself as free: but it is grossly mistaken: it is free only during the election of members of parliament. As soon as they are elected, slavery overtakes it, and it is nothing."[11]

The Roman cult of *libertas,* or autonomy to use the language of modern liberalism, is structured around personal rights, not around political freedom. The distinction between autonomy and freedom is by no means trivial: the "autonomy" of the individual is structured in theory around the Roman and liberalistic notion of a seemingly sovereign, self-contained individual who has no clear roots in social life, while the word *freedom* (derived from the Germanic *Freiheit*) implies that individuality has deep social roots and responsibilities. In imperial Rome, particularly under the Julio-Claudian emperors, the individual could exercise a wide range of choices in vocations, responsibilities, and the satisfaction of tastes. Roman license during the Empire, with its almost psychotic appetite for extremes, merely expanded this cult into a way of life. Credo was warped into extravagant practice with the result that the state soon found it ceased to enjoy the support of its citizenry. Its citizens fled from military service, public obligations, tax levies, and the most minimal communal responsibilities. Accordingly, every aspect of government had to be professionalized. Under the Empire, Rome's troops were mercenaries, increasingly of alien birth and culture; its bureaucracy became an elaborate apparatus, staffed by numerous ex-slaves who had acquired the skills for political affairs that their former masters lost and that Roman citizens generally neglected.

This extraordinary erosion of personal competence blemishes every aspect of the imperial era. But even the republic prepared the way for Rome's decline. In the political sphere, Roman *libertas* never became freedom, the Greek vision of *eleutheria* based on equality. One searches Latin for a term other than *libertas* or *licentia* that expresses the centricity of individual political judgments, in short, a term that does not sort the individual into the collective and weighted units of the *comitia* and *councilium*. The search is a vain one. The Roman concept of political life is corporatist, even statist, to the core, and there is no reciprocal interaction between the personal and the political.

To find an individual who has room for a political life, we turn to the Roman noble or well-born for examples. Here, political life is obligatory—indeed, apart from war, the authentic calling of an aristocrat. Hence it is to be conceived as a profession and suffers from the very professionalization of politics that Greek democrats tried so assiduously to avoid. Young men of patrician lineage were trained from birth in diction and rhetoric, physical fitness and military skills, amiability and the arts of influence. Overly mannered and self-conscious, they were taught to gain favor with the powerful and befriend the potentially influential. By degrees, they were initiated into legal skills and affairs together with martial arts and the postures of command. Polybius advises us that ten years of military service are necessary before a man can aspire to a political career—a prerequisite that was mercifully abbreviated in the later years of the Republic. One then went into the service of a provincial governor and moved onward, at home, to a minor magistrate or a military tribune.

To become a praetor was the "next obligatory office," as Taylor puts it, followed by the consulship, if at all possible. Between times, one held important offices in the provinces where the opportunities for enrichment and plunder were immense. These broad outlines of the nobleman's training and career could be painted with details that more appropriately describe politicking, rather than politics. The right friends, devoted clients, and suitable personal connections were critically important in achieving public office and political renown.

Even more significant than the Roman cult of *libertas* was the Roman cult of *amicitia*, the Latin word for "friendship." Career success depended not only upon lineage and wealth but also on the elaborate system of friends and shared obligations a rising patrician developed. I use the word "system" advisedly to single out the complex machinery of personal ties and interactions on which the whole structure of rule was based. Roman "politics" must be seen as a network of clients and associates rather than clubs and parties. The fierce differences between factions in the Republic that finally brought it to ruin were more personal in nature than political.

Cicero's allusions to the *partes* of the *populares* and the *optimates* (the "parties" of the "people" and "aristocrats," to use these terms in a modern sense) are evidence of differences in methods of manipulation rather than programs. None of the nobles, with the exception of the brothers Gracchi, ever tried to really shift political power from the patrician elite to the populace. Indeed, Roman politicians were rarely burdened by sentimentality for the oppressed or the plight of the commoners.

To use the people for personal ends and career ambitions, however, was a widespread technique, not only during republican times but also during the imperial era. Nor were the Roman people the worse for the use of such demagogic tactics. Nobles gave immense quantities of their wealth to gain popular support against their rivals. A steady flow of emoluments, gifts, festivals, and games came to be expected by the Roman people as a characteristic feature of politics. Roman client and gift politics, in turn, accelerated the degradation of the citizenry, fostering an appetite for sensationalism and brutality that emerged in marked contrast with its traditional republican spirit and virtues. "Public things" or *res publica* became a highly merchandisable commodity—a "thing" to be sold, bought, and pilfered. In this respect, the Empire changed very little in Roman "politics." It merely made the process of demoralization, vulgarization, and pilfering more systematic and orderly.

Early Rome did not produce a breed of kindly men—nor, for that matter, did Athens and other Greek *poleis*. Kindness and sentimentality are not classical traits. Obligation and duty are the preferred personal attributes of the ancient world. But these attributes did create an ideal of a highly committed, morally certain, and fiercely independent yeomanry. The landholdings of these yeomen provided the material competence for a solid independence of mind and a sense of community rootedness. Behind the more distinguished names of early republican Rome, such as Cincinnatus, who left his farmstead for vigorous public service, were the stern traditions of family cults, civic deities, and an unblemished ancestry—a lineage to be cherished because it exhibited soldierly simplicity and agrarian virtues. *Dignatas* and *honorare* were to be prized over wealth, social status, and public esteem, although invariably such rewards came with family probity.

These stern and dutiful farmers were to fill the legions commanded by Scipio in the brutal wars with Carthage. They were the fodder of costly, long-term, and debilitating conflicts that brought ruin to their farms and the destruction of their social moorings. Thereafter, an unyielding patriciate, too urbane to value the innocent simplicity of its own rural ancestors, effaced what the Punic Wars had largely undone—the ideals of republican virtue and the agrarian material conditions in which this sense of virtue and duty was rooted. Cincinnatus belonged to a social world rather than a political one. Governance in his day was seen as a domestic responsibility in which a public servant tended to the needs of the people more as a father than an administrator. Such men became short-lived *dictatori* without ever establishing dictatorships; they promptly went back to their farms after answering the call to public service. They did not thirst for power, much less professionalize it.

But Rome could not strike the balance between aristocratic values and public rights achieved by Athens. More precisely, Rome failed to turn the governance of the *civitas* into a genuinely political community. Men like Cincinnatus were to lose not only the landholdings that gave them independence of mind and spirit; they were to lose the social base for public commitment without developing a politics that could control and contain the new civic dispensation that was forming around them. Like Athens, Rome was to grow—and, like Athens, it was to be brought into a broad regional theater of power relationships and responsibilities. But where Athens drastically reworked its yeoman society into a vital public realm that fostered active citizenship among all its social elements, Rome permitted its yeomanry to dissolve into rootless constituencies and its public life to languish. A republic rather than a democracy came into existence with a degree of administrative ingenuity unprecedented in the history of jurisprudence, efficiency, and military prowess. But for this achievement it paid a penalty that ultimately spelled its death. The late Roman Republic was not a world that could nourish a Cincinnatus or even a Marcus Portius Cato, whose writings are filled with denunciations of Rome's moral debasement, lasciviousness, and extravagance. Politics was claimed almost exclusively by the patriciate and jealously guarded from any serious invasion by the people. In this sense, republican Rome was true to itself: like all elitist regimes, it would have been exceptional if it failed to turn from an increasingly oligarchical republic into a completely despotic empire.

Athens and Rome ultimately became legendary models for two types of "popular" government: a democracy and a republic. And later social theorists and political practitioners who had lost any monarchic proclivities were to clearly favor a republican system of governance over a democratic one.

But democratic notions of a body politic did not disappear. They were to surface from the depths of a popular "underground" of deviant Christian sects throughout the Middle Ages, such as the Brethren of the Free Spirit, Anabaptist movements, and blatantly anarchic conventicles during the Reformation Era. Like Athens, they were not without their flaws: elites are to be found within elites, saints within larger communities of believers, and the like. Nevertheless, village democracies kept alive strong traditions of popular assemblies that may have been inherited from distant Neolithic and similar institutions and that also emerged in many medieval towns. The notion of the "people" or *demos* did not disappear. The ideals of "popular rule" were to linger on from classical antiquity well into modern times.

More commonly, however, republican theories of governance were hybridized with democratic notions, and they were to produce rising demands of self-governance with institutions that were redolent of democratic Athens. Machiavelli's *Prince* and *Discourses* glitter with a fascinating mixture of republican and democratic ideas, largely translated into the virtues of his beloved Roman Republic. His aversion for the idle nobility is pronounced. Politics, in Machiavelli's eyes, is not the fare for the slothful, ignorant, and crested boors who are the fatuous heirs of titles and aristocratic pedigrees. It is a highly skilled craft that must be exercised as gently by the prince in his relationship with the people as it must be exercised ruthlessly by him in his relationship with his rivals. Machiavelli's demand for a total commitment by his chosen sovereign to politics reflects the emergence of a new kind of man, the Renaissance prince: secular, keenly intelligent, skillful, and cunning. He is a man of reason rather than faith, of judgment rather than belief, and self-reliance rather than dependency. A new political dispensation is in the air, a modern one, that draws its precedents from the senatorial party of the early Roman Republic rather than the sacerdotal party of the medieval church. Machiavelli's references are to Scipio, not to Augustine; to Livy, not Aquinas.

But within this republican idea of a meritocracy, Machiavelli advances sixteenth-century Italian concepts that could be found in Perikles and the Athenian commitment to amateurism. Comparable only to his hatred of the nobility is his hatred of mercenaries who were plaguing and plundering Italy in his day—professional soldiers. The most commanding need of a well-ordered state, he tells us, is a citizen-army. Mercenaries are as unreliable as they are unscrupulous. They are born plunderers who have no allegiances other than those that money can buy. "Mercenary captains are either very capable men or not," he declares; "if they are, you cannot rely on them, for they will always aspire to their own greatness, either by oppressing you, their master, or by oppressing others against your intentions; but if the captain is not an able man, he will generally ruin you. And if it is replied to this, that whoever has armed forces will do the same, whether these are mercenaries or not, I would reply that as armies are to be used either by a prince or by a republic, the prince must go in person to take the position of captain, and the republic must send its own citizens. If the man sent turns out to be incompetent, it must change him; and if capable, keep him by law from going beyond the proper limits. And it is seen by experience that only princes and armed republics make very great progress, whereas mercenary forces do nothing but harm, and also an armed republic submits less easily to the rule of one of its citizens than a republic armed by foreign forces."[12]

Machiavelli's argument clearly tips toward a republic and an armed citizenry rather than a prince and a professional army.

Clearly, if princely government was central to his concerns, the prince's competence would normally be beyond any legal assessment. There would be nothing that could prevent him from "going beyond the proper limits," indeed, to tolerate any limits at all. Nor could Machiavelli, whose mind was steeped in the Greek and Roman classics, have been unmindful that the Athenian military forces, in contrast to the Roman imperial ones, were structured around clearly accountable captains who were strictly regulated by law, indeed by the *ekklesia* or popular citizen-assembly. Amateurism takes high priority over professionalism and political institutions, visibly peopled by a free, and an armed, citizenry, over a state power with its mercenary bureaucrats and soldiers.

Machiavelli undoubtedly had his eyes sharply focused on Italy and the cause of national independence, but his feet were firmly planted in his beloved Florence and the cause of freedom. The sap flowed, as it were, from the roots upward—from city to nation—with the result that a republican, even princely, state was nourished by and mixed with civic democratic notions.

Ideologically, the hybridization of two very distinct and potentially conflicting classical ideas of public governance arises from a serious confusion over what we mean by politics and statecraft.

The rise of the nation-state from the sixteenth century onward greatly altered the entire framework of political discourse. The basic unit of public governance was the city, not larger entities such as the province, nation, or empire. A citizen's allegiances to governing institutions could be comfortably enumerated as a very distinct hierarchy of loyalties. He was first and foremost a townsman. The town was the authentic and most meaningful locus of his personal and public life. Only secondarily did he identify himself with a province or a region. The idea of "nationality" was at best vague, that is when it existed at all. Romans in the most far-flung reaches of the empire consistently visualized themselves as citizens of the imperial city. Among the Greeks, civic loyalty was virtually all-consuming: Athenians, for example, sharply and disdainfully distinguished themselves from

Spartans or Corinthians, a sentiment that was freely reciprocated by all citizens of other *poleis* with respect to other Greek cities. The Stoic philosophers who were to pave the way for Christianity insisted well into Roman times that civic loyalty defiled the novel notion that all men were brothers. But the Stoic notion and its very novelty has the ring of an ideological protest against the more popular view that citizenship implies a primary loyalty to one's town, not to a vagary called "humanity."

The rise of the nation-state altered this hierarchy of loyalties—and, with this change, the way in which politics was conceived. Not that ancient and medieval civic parochialism was an unblemished desideratum. Parochialism had a very harmful, often dehumanizing, effect on urban life generally. The tendency to set one city against another fostered local chauvinism with such pathologies as ethnic antagonisms, wars, and cultural introversion. In a world where the city produced a deep sense of ethnic and cultural identity that compares with the modern world's most strident forms of nationalism, the conquest of one city by another often terminated in the sheer annihilation of a people as a distinct community. Rome's total destruction of Carthage in the last of the Punic wars was not merely the dismantling of a major ancient city; it was the enslavement and total effacement of a people—of their identity, culture, traditions, uniqueness, indeed their very claim to exist. Jericho, Troy, and Jerusalem were to suffer similar fates, to cite only the most well-known examples of what urban destruction often meant in the early and classical worlds—an act of devastation comparable only to genocide in the modern world.

With the rise of nationalism and the nation-state, the state began to assume ideological preeminence over the city, and even radical social thinkers began to formulate their political ideologies in broad territorial or national terms. Puritan revolutionaries thought of their "rights" not as citizens of London, which formed the real center of parliamentary unrest against the court, but as "Englishmen." Puritan theory, based on a doctrine of natural rights, formulated these "rights" not in the characteristically civic forms of popular assemblies and politics based on personal intimacy; a nation, "England," was conceived as the legendary victim of invading "barbarous" Normans who had crossed the channel from France some five hundred years earlier and imposed a royal tyranny on a representative system of Saxon self-governance. The combatants in the revolution expressed their loyalties in terms of their adherence to "parliament" or the "court." Larger-than-life institutions, far beyond the reach of the ordinary citizen, began to supplant the civic institutions within which some kind of face-to-face democracy was feasible. Republicanism, in effect, was a radical ideology of nations rather than cities and statecraft became the "politics" of highly centralistic state structures.

It is hard to overstate the amount of intellectual mischief the extension of the word politics, basically rooted in the civic life of the *polis*, produced when it was permitted to encompass statecraft. Classical politics always implied the existence of a body politic—in its own way, a kind of ecological community in the social sense even in Rome, when the words *populus Romanus* came to mean little more than an aristocratic oligarchy. The classical notion of a body politic was not a euphemism for an "electorate" or a "constituency," as it is today; it was a real, physical, and clearly observable entity. It could be seen daily in public squares where heated discussions over political issues intermingled with the chitchat of personal and business problems; it assembled with almost weekly regularity on a hillside of Athens, the Pnyx, where meetings of the *ekklesia* were convened, or it gathered in open spaces of the Roman Forum where the *comitia tributa* often held its sessions. It could be heard quite audibly, whether by acclamation in Sparta, arguments in Athens, or even in the most despotic of Rome's imperial periods when the hoots and shouts of commoners at the Coliseum reminded the emperors that they were not beyond the reach of public criticism. In more militant times, this body politic rioted in Rome's St. Peter's Square during the Middle Ages and stormed into Florentine churches to hear the sermons of Savonarola during the Renaissance. In short, the body politic existed in the literal sense that it was a tangible, protoplasmic entity that expressed its concerns in the eye-to-eye contact of personal confrontation and fervent discourse.

This eye-to-eye contact of active citizens was an organic politics in its most meaningful, protoplasmic, and self-fulfilling sense. Political assemblies were not mere audiences on which public officials practiced their arts of statecraft; they were legislative communities united by a reasonable commonality of shared, public interests and ethical precepts. That political life had worked its way out of social life to acquire a distinct identity of its own and presupposed social forms as its underpinnings is evident enough from any account of the *polis* or its near-equivalent in the medieval city-state or "commune." But even so conservative a thinker as Aristotle never confused a family or a workshop with the *agora*, where public affairs were normally discussed, and the *ekklesia*, where the body politic physically assembled to make public decisions. Hence, the Greek *polis* was never a state in any modern sense of the term with professional surrogates for an assembled body politic, nor was it a social entity such as a "family" that united the people into an authentic kin group. Aristotle's notion of *philia* or solidarity as a crucial precondition for a political life expressed the unique identity politics possessed as a form of governance, one that transcended mere kinship obligations. If kinsmen were obligated to each other by virtue of blood ties and tribal custom, citizens were obligated to each other by virtue of civic ties and ethical precepts.

If politics can be said to have emerged from society in the strict sense that I use the latter word to denote familial, vocational, and sociable relationships, so statecraft can be said to have emerged from politics conceived as the activities of a directly involved body politic. Aristocracies, monarchies, and republics ultimately dissolve the body politic as a participatory entity, an essentially ecological phenomenon into an amorphous mass of privatized "social" beings we so aptly call an electorate or a constituency. The "deputies of the people" replace the people, to use Rousseau's pithy formulation, and bureaucratic institutions replace popular assemblies. The identity of politics as a unique phenomenon to be distinguished from other, presumably "social" activities, is not a concept that was confined to classical thinkers such as Aristotle. It is a recurring and often perplexing problem that appears in the writings of Rousseau, in constitutional documents of the past that distinguish between "active" (propertied) citizens and "passive" (propertyless) citizens, and today, most strikingly, in the writings of a highly gifted political philosopher, Hannah Arendt.

What is so curious about this literature and its attempt to single out politics as a clearly identifiable area of public activity is the extent to which it is burdened by the institutional weight of the nation-state. Arendt's distinction between a "political realm" and one that is "social" allows for very little difference between political activity and statecraft. The state has so thoroughly merged with the political—institutionally and functionally—that the two almost seem identical. What is remarkable is that modern social theory does not find this congruence of very different arenas of public governance problematical. Clear as the old Aristotelian distinction between the social and political may be, the equally crucial distinction between the political and the statist tends to be lost in the modern literature on politics. Political activity and statecraft have become so thoroughly intermixed in theory and reality that the present-day usage of the word "politics" is taken to be the "art" of the politician, who, for all practical purposes, replaces the body politic. That the state historically depoliticized this body politic and essentially disbanded it institutionally seems like a meaningless ideological curiosity in a world where "political activity" takes the form of an on-going battle of political gladiators in a strangely muted, almost empty arena.

Perhaps the main reason why the confusion between politics and statecraft persists so strongly today is that we have lost sight of the historic source and principal arena of any authentic politics—the city. We not only confuse urbanization with citification, but we have literally dropped the city out of the history of ideas—both in terms of the way it explains the present human condition and the systems of public governance it creates. Not that we lack any valuable histories of the city or attempts to evaluate it sociologically. But our urban literature generally neglects the relationship between the city and the remarkable phenomenon of citizenship it produces. Urban historians tend to fixate on largely narrative accounts of the city's development from village to megalopolis—accounts that are riddled by nostalgia for the past or a brute acceptance of existing urban conditions and the future they portend. The notion that the city is the source of immensely provocative political, ethical, and economic theories—indeed, that its institutions and structures embody them—is generally alien to the modern social theorist.

An ethical interpretation of historical urban standards must highlight one central issue: the need to recover civic forms and values that foster an active citizenry. This amounts to saying that we must recover politics again—not only the social forms of personal intercourse that underpin every kind of human activity. The city, conceived as a new kind of ethical union, a humanly scaled form of personal empowerment, a participatory, even ecological system of decision making, and a distinctive source of civic culture—this civic notion of community must be brought back again into the history of human ideas and practical wisdom. It must be critically reexamined as a realm of thought and activity that gives rise—as it did in various periods of history—to political consociation, a politics that places family, work, friendship, art, and values within the larger context of a rounded civic world. Politics, in effect, must be recreated again if we are to reclaim any degree of personal and collective sovereignty over our destiny. The nuclear unit of this politics is not the impersonal bureaucrat, the professional politician, the party functionary, or even the urban resident in all the splendor of his or her civic anonymity. It is the citizen—a term that embodies the classical ideals of *philia*, autonomy, rationality, and, above all, civic commitment. The elusive citizen who surfaced historically in the assemblies of Greece, in the communes of medieval Europe, in the town meetings of New England, and in the revolutionary sections of Paris must be brought to the foreground of political theory. For without his or her presence and without a clear understanding of his or her genesis, development, and potentialities, any discussion of the city is likely to become anemically institutional and formal. A city would almost certainly become a shapeless blob, a mere chaos of structures, streets, and squares if it lacked the institutions and forms appropriate to the development of an active citizenry. But without the citizens to occupy these institutions and fill these forms, we may create an endless variety of civic entities—but like the great urban belts that threaten to devour them, they would be completely socially lifeless and ecologically denatured.

Chapter Four
The Ideal of Citizenship

If the city makes it possible for us to single out politics as a unique sphere of self-governance that is neither social nor statist, the citizen as the viable substance of this unique sphere makes it possible to undo the confusion that blends these very distinct spheres into a collage of overlapping terms and blurred meanings. For it is in the citizen—in his or her activity as a self-governing being—that the political sphere becomes a living reality with the flesh and blood of a palpable body politic.

The Greeks may have been the first people to give us a clear image of the citizen in any politically intelligible sense of the term. Tribal peoples form social groups—families, clans, personal and community alliances, sororal and fraternal clubs, vocational and totemic societies, and the like. They may assemble regularly to examine and decide communal affairs—certainly a nascent form of politics—but the issues that confront them rarely deal with ways and means of governing themselves. Custom plays a paramount role in establishing their norms for community management; discourse, beyond direct argumentation, occupies a place secondary to the enormous authority of precedence and long-established administrative procedures. Nor is this approach to be disdained as trivial or "primitive." Group safety and stability require that the community preserve the old, well-tested ways of life, of expeditiously applying and modifying time-honored and secure structures of group management. The kin relationship forms the social tissue of this governing body, whether the blood tie be real or fictitious. Religious belief, too, may play a very important role, as Fustel de Coulanges has argued.

But politics as a creative and rational arena of discourse with its vastly innovative possibilities for shaping and bonding widely disparate individuals is only latent in tribal assemblies. It still has a domestic character with powerful familial biases that exclude the stranger. Tribal assemblies of preliterate peoples invoke the past; political assemblies of free citizens create a future. The former tends to be highly conservative; the latter, highly innovative. If the two were juxtaposed with each other, we would be obliged to contrast custom to reason, precedent to a sense of futurity, kinship ties to civic ties, mythopoeia to ethics. In waxing enthusiastically over the popular assemblies that existed very early in Mesopotamian cities, Henri Frankfort declared that the assembly form "is a man-made institution overriding the natural and primordial division of society into families and clans. It asserts that habitat, not kinship, determines one's affinities. The city, moreover, does not recognize outside authority. It may be subjected by a neighbour or a ruler, but its loyalty cannot be won by force, for its sovereignty rests with the assembly of its citizens."[1]

Which is not to say that these contrasts are so absolute that they polarize the tribal assembly against its civic counterpart. Athenian citizenship, based on a civic myth that all citizens shared a common ancestry, became highly parochial by Perikles's time. For a well-established resident alien or *metoikos* (metics) to become a citizen of the *polis* was virtually impossible. Doubtless, Athenians knew that Solon, a century earlier, offered the lure of Athenian citizenship to all skilled craftsmen who were willing to migrate from various parts of the Mediterranean to Athens. And Kleisthenes, a generation removed from Perikles, permitted many metics to become citizens in his day. Athenian citizenship, conceived as a form of status based on blood ties, was a shaky affair at best. But under Perikles, the body politic behaved like an oversized medieval guild. For patently self-serving reasons, it simply closed its doors to outsiders who might stake a reasonable claim to the privileges afforded by the corporate community. In principle, the impediments Athenians raised to citizenship in the middle of the fifth century B.c. were not different from those that modern nation states place in the way of immigrants and undocumented residents.

The Greek citizen ideal, however, differed very profoundly from the modern. It was not simply some specious myth of shared heredity that united citizens of the *polis* with each other but a profoundly cultural conception of personal development—the Greek notion of *paideia*. *Paideia* is normally translated into English as education, a term that is notable for its sparseness and limitations. To the Greeks, particularly the Athenians, the word meant considerably more. The education of a young man involved a deeply formative and life-long process whose end result made him an asset to the *polis*, to his friends and family, and induced him to live up to the community's highest ethical ideals. The German word, *bildung*, with its combined meanings of character development, growth, enculturation, and a well-rounded education in knowledge and skills, more appropriately denotes what the Greeks meant by *paideia* than any word we have in English. It expresses a creative integration of the individual into his environment, a balance that demands a critical mind with a wide-ranging sense of duty. The Greek word, *areté*, which in Homeric times denoted the warrior attributes of prowess and valor, was extended by the classical era to mean goodness, virtue, and excellence in all aspects of life. *Paideia* and *areté* are indissolubly linked—not as means and ends but as a unified process of civic- and self-development. Excellence in public life was as crucial to an Athenian's character development as excellence in his personal life. The *polis* was not only a treasured end in itself; it was the "school" in which the citizen's highest virtues were formed and found expression. Politics, in turn, was not only concerned with administering the affairs of the *polis* but also with educating the citizen as a public being who developed the competence to act in the public interest. *Paideia*, in effect, was a form of civic schooling as well as personal training. It rooted civic commitment in independence of mind, *philia*, and a deep sense of individual responsibility.

The modern notion of "politics" as a form of managerial "effi-ciency" or of education as the mere acquisition of knowledge and skills would have seemed pitiful to an Athenian citizen of classical times. Athenians assembled as an *ekklesia* not only to formulate pol-icies and make judgments; they came together to mutually educate each other in the ability to act justly and expand their civic ideals of right and wrong. The "political process," to use a modern cliche, was not strictly institutional and administrative; it was intensely processual in the sense that politics was an inexhaustible, everyday "curriculum" for intellectual, ethical, and personal, growth—*paideia* that fostered this ability of citizens to creatively participate in pub-lic affairs, to bring their best abilities to the service of the *polis* and its needs, to intelligently manage their private affairs in accordance with the highest ethical standards of the community.

This "calling" to civic and personal excellence was more than a family responsibility or an institutionalized form of personal train-ing. By classical times, Athenians who could afford them had tutors aplenty—rhetoricians to teach them the arts of persuasion; philoso-phers and logicians to instruct them in wisdom and consistency of thought; elders to provide them with the inherited lore of their fam-ilies, civic traditions, and models of behavior; gymnasia in which to train and control their bodies or learn martial arts; courts to shape their faculties for judgment; and, in time, the *ekklesia* in which to for-mulate crucial policies through discourse and debate. But every *polis*, be it a garrison-state such as Sparta or a democracy such as Athens, provided a variety of public spaces in which citizens could gather on more intimate terms, often daily, to discuss public and practical af-fairs. Perhaps the most important of these spaces, the *agora*, which M. I. Finley calls the "town square," was an informal meeting ground in which the people could be assembled when needed.[2] By Perikles's time, the Athenians were to shift the formal assembly of the people—the *ekklesia*—to a hillside (the *Pnyx*), but, as Finley notes, the *agora* originally meant a "gathering place," long before it was invaded by shops, stalls, and temples.

The *agora* provided the indispensable physical space for turning citizenship from a periodic institutional ritual into a living, everyday practice. Home was the place in which one ate, slept, and tended to the details of private life. But the Greeks generally held this private world in small esteem. Life was authentically lived in the open public space of the *agora*, where the citizens discussed business affairs, gossiped, met friends—new and old—occasionally philosophized, and almost certainly engaged in vigorous political discussion. Perikles could be waylaid there by badgering critics as surely as Sokrates could be drawn into lengthy discussions by the intellectually earnest young nobles of the *polis*. Jugglers, acrobats, poets, and play-actors mixed with tradesmen, yeomen, philosophers, and public officials, a crowd spiced by strangely costumed visiting foreigners who gawked at the looming acropolis above and the superbly adorned public buildings nearby.

During inclement weather, this colorful and eminently vocal crowd could take refuge in the colonnaded arcades or *stoa* that lined part of the twenty-six-acre square. There, they encountered artisans working at their trade and merchants who displayed their wares, often women who sold much of the farm produce that fed the community. In its emphasis on direct, almost protoplasmic contact, full participatory involvement and its delight in variety and diversity, there is a sense in which the *agora* formed the space for a genuine ecological community within the *polis* itself. Thus politics, which found its most ordered and institutionalized expression in the *ekklesia*, originated in the daily ferment of ordinary life in the *agora*. Its informal genesis reveals the organic way in which important policies slowly developed into popular ideas before they were formulated as verdicts and laws in the courts and official assemblages of the *polis*. The democratic institutions of Athens, for all the ritualistic panoply that surrounded them, were merely the structural forms in which everyday debate and gossip were hardened into the legislated expression of an easy-going, unstructured, and popular politics—one that was embodied by an earnest, spontaneous, and an extraordinarily active citizenry. The "tyranny of structurelessness" that so many contemporary liberals and socialists fling so reprovingly at their libertarian critics as an "ultrademocratic" vice would have been incomprehensible to the Athenian citizen. The teeming "anarchy" of the *agora* was, in fact, an indispensable and fecund grounding for "libertarian" structures (to use Jaeger's term) that, given time and neglect, would have otherwise turned into oligarchic institutions with a democratic veneer.

Citizenship, in effect, involved an on-going process of educational, ethical, and political gestation for which such words as "constituent" and "voter" are modern parodies of a politics that was more existential than formal. This gestative process occurred in the *agora* days before it found expression in the *ekklesia*. It is a cliche to say that Americans are a "practical people," Italians an "emotional people," Germans a "methodical people." In any case, there can be no doubt that the Athenians were a "political people" and citizenship was their destiny, not merely the avocation it has become in the modern world. Behind their ideal of citizenship, nonperishable as long as there is a meaningful literature on democracy, is the way that citizenship was formed. It was a citizenship formed by the moral fortitude of a mountain people—in Athens's case, tempered by the wide cultural contacts afforded by a major seaport—and a delicate balancing of social conflicts that pitted opposing classes against each other without obliterating the virtues each possessed in its own right. Greece was to inherit not only the aristocratic epics of Homer and their high standard of courage but the mundane litanies of Hesiod with their workaday sense of practicality. Her long journey from a tribal to a political world, from a society of peasants to one of citizens, is a fascinating narrative in its own right. But it is also an exemplary biography of the citizen as such, at least insofar as this remarkable individual was to approximate an ideal that Athens more closely attained than any community that followed her—and one that the modern world may well lose at the peril of its freedom.

The first phase of this journey into a political world begins with the way the Athenians managed to shed the narrow features of the kinship bond based on blood ties, religion, and familial loyalties, while simultaneously developing an almost ecological *sensibility* based on a fictive kinship, territorial commonality, rationality, and a healthy secular humanism.

Greek society was not immune to a general historical trend that raised humanity out of a dismal archaic era, one which reworked an egalitarian tribal world into a hierarchical feudal one. By Homeric times, commanding patriarchal clans had already imposed their will on loosely structured and highly vulnerable peasant communities. Ironically, the very tribal features of kin and blood that once had produced egalitarian norms of mutual aid and material reciprocity were used to achieve their very opposite: a hierarchical system of rule focused on domination and acquisition. This new reordering of traditional clans or phratries from tribal into aristocratic families fostered exclusivity and privilege rather than sharing and communal responsibility. The "bribe-devouring judges" whom Hesiod, the Boeotian peasant, had denounced in the eighth century B.C. had become land-devouring nobles by the seventh. Gross inequalities in the ownership of land that mark the Greek world in that period of transition seem like forerunners of the Roman agrarian crisis that followed the Punic Wars. That Greece, especially Athens, did not become a mere historical preface to Rome, indeed that the *polis* developed a political life so markedly different from any republican system of governance, can be explained only in terms of a remarkable constellation of factors from which the modern world can learn much.

What aristocratic Greece and democratic Greece were to share as a common legacy is a vigorous ideal of independence—an ideal even stronger than its widely touted ideal of justice. To the Greek mind, clientage in any form verged on slavery, indeed a denial of the individual's humanness and personality. This notion, which many Greek scholars were to regard as an aristocratic disdain for work as such, actually expressed a concern for the citizen's capacity to form independent judgments insulated from external or personal interests. "To build one's own house, one's own ship, or to spin and weave the material which is used to clothe the members of one's own household is in no way shameful," observes Claude Mossé in her insightful study on work in the ancient world. "But to work for another man, in return for a wage of any kind, is degrading. It is this which distinguishes the ancient mentality from a modern which would have no hesitation in placing the independent artisan above the wage-earner. But, for the ancients, there is really no difference between the artisan who sells his own products and the workman who hires out his services. Both work to satisfy the needs of others, not their own. They depend on others for their livelihood. For that reason they are no longer free. This perhaps above all is what distinguishes the artisan from the peasant. The peasant is so much closer to the ideal of self-sufficiency (*autarkeia*) which was the essential basis for man's freedom in the ancient world. Needless to say, in the classical age, in both Greece and Rome, this ideal of self-sufficiency had long since given way to a system of organized trade. However, the archaic mentality endured, and this explains not only the scorn felt for the artisan, labouring in his smithy, or beneath the scorching sun on building sites, but also the scarcely veiled disdain felt for merchants or for rich entrepreneurs who live off the labor of their slaves."[3]

As befitted free men, farmers enjoyed the economic independence and material security that were needed to form decisions untainted by self or class interests. In fact, many Greeks would have seen even wealthy tradesmen as clients of their buyers and highly skilled craftsmen, artists, and poets as dependents of a fickle market for their products. To be free in Athens meant very little if one's basic needs were not satisfied within a mutualistic group of self-sufficient producers. The word *autarkeia*, strictly translated, has the double meaning of the rarely used definition of "self-rule," as well as the more familiar notion of "self-sufficiency." In its latter meaning, *autarkeia* has long been replaced by *autonomos*, literally the condition of living by one's own laws. This concept of independence is more juridical than political. The English translation of these words, notably the use of "autarchy" to mean economic self-sufficiency and "autonomy" to denote personal freedom or self-government, creates a disjunction between the material and political that would have been alien to the Greek ideal of independence. We would be hard put to understand Aristotle's belief that all tradesmen, artisans, merchants, and servants should be denied the franchise if we failed to recognize that it is not simply labor and trade he despised but, more importantly, material clientage in any form that could affect the citizen's independence of judgment. Independence without the substance of material self-sufficiency and personal autonomy would have been formal at best and hollow at worst to the Greek mind. No client, however well-off, could render a judgment or reason freely without deferring to exogenous authorities and interests on whom his welfare depended.

It is worth noting that these Hellenic precepts have entered into modern ecological thinking with little knowledge of their remote origins and political orientation. But such a demanding notion of independence could not have emerged solely from ideological considerations. Here, the mountainous terrain of the Greek archipelago comes very much to our aid. "Nature gave to Greece, as to her neighbors, the tendency to equality together with abundant opportunities for the growth of public opinion, and then intensified these forces by strictly limiting the areas in which they could operate," observes Alfred Zimmern. "Each little plain, rigidly sealed within its mountain-barriers and with its population concentrated upon its small portion of good soil, seems formed to be a complete world of its own. Make your way up the pasture-land, over the pass and down on to the fields and orchards on the other side, and you will find new traditions and customs, new laws and new gods, and most probably a new dialect.... The Greeks were not painfully taught to value local independence. They grew up unable to conceive of any other state of government. It was a legacy slowly deposited through the long period of isolation which intervened between the first settlement of the Hellenic invaders and their emergence centuries later as a civilized race. They never themselves realized, even their greatest writers did not realize, how unique and remarkable their political institutions were."[4]

Colorful and truthful as these lines may be, Zimmern understates the extent to which Greek thinkers were conscious of their heritage and political uniqueness. Indeed, as we shall see, it was the extraordinary acuity of this consciousness that imbued Athens with a high sense of mission and clear sense of direction. No political sphere was more carefully, thoughtfully, and artistically crafted than the Athenian. Equality or *isonomia* followed from an even more basic influence than Greece's mountainous terrain exercised on its pocket-sized communities of free villagers. To live in such isolation and forced independence required that every family and family-sized community had to fend for itself. The Greek farmer had to be a well-rounded man in a more tangible sense than we use this cliché today. He was obliged to know how to fight for his land as well as cultivate it, build his shelters as well as repair them, methodically tend to his wounds as well as cunningly avoid needless conflicts, plan his long-range needs as well as satisfy his immediate ones, function as a caring father, loyal spouse, dutiful son, supportive brother, a wary buyer of things that his community could not produce—in short, he needed to combine a working knowledge of all the techniques needed for his survival with soldiering and "politicking." Together with his spouse, who presided over all the domestic affairs of the family, the Greek farmer gained in practical competence and personal fortitude what isolation denied him in acculturation. In the Latin sense of *competere*—to be fit, proper, or qualified—he had no equal in the ancient world. Hence, he could be "unequalled" by self-anointed superiors who tried to subordinate him and assert their authority over his destiny. A magnificent amateur, he embodied the nascent citizenship in which all his peers acknowledged the need for a self-possessed individual who could be entrusted as much with the affairs of his community as with the satisfaction of his private needs.

The flow of ideas from the independence of a mountain-dwelling villager to the egalitarianism of a *polis*-dwelling citizen must be seen as an unbroken continuum. The Greek language itself is magnificently processual and organic. A crucially important word such as *arche*, from which an entire political vocabulary has been constructed, denotes the originating principle as well as the ordering principle of any *kosmos*, the Greek word for "order" in the broadest sense of the term. To think of an "order" without deriving it from its origin and, hence, the latent possibilities that it could fulfill is linguistically built into the Greek mind—and also provides us with an important clue to the language's ethical thrust. Origin, history, fulfillment, and possibility—all form a unified whole in Greek, so that whether we choose to speak of the universe, humankind, the *polis*, or the citizen, these concepts, charged with ethical meaning, denote the unity of civic, natural, and social life or what we call "connectedness" today. Accordingly, autarchies or self-sufficient communities lead us organically—deductively, if you will—to independence, competence, and *isonomia*. The Greek *polis* has its *arche* in this germinal phasing of a highly competent farmer who, by an immanent process of sociopolitical development, found his fulfillment as a highly competent citizen.

The old Greek aristocracy was no mere anachronism in this process. The Athenian democracy did not shed it; rather, it tried, with qualified success, to absorb it. Its epic culture, gospel of valor, high sense of *philia*, and code of honor, marked by a disdain for material things, were incorporated into the puritanical virtues of the democracy, which abjured luxury, ornateness of dress, culinary delights, and self-indulgence. The democratic hero was not only valorous but sternly self-willed and emotionally controlled. Warrior manliness did not die; it was reworked into civic loyalty, personal dignity, and a high regard for virtuous behavior. The comradeship of the military camp became the *isonomia* of the *ekklesia* in which human worth was seen as virtually interchangeable. Citizens were expected not only to be competent but to be competent equals. Hence, all could participate in the governance of the *polis* on the same footing, a practice which, as we shall see, was translated into the widespread use of sortition and attempts to arrive at decisions by consensus as well as by voting.

It is worth emphasizing that nearly all the men who turned Athens from an aristocratic oligarchy into a democratic polity were of noble lineage. Solon, Kleisthenes, and Perikles, to cite the most well-known figures in forming the democracy, were members of the most elite clans, the *genos*, of Attica, the territorial domain of Athens. In contrast to their Roman counterparts, however, they did not become the leaders of an unruly *partes populares*, nor did they try to rise in an oppressive hierarchy by throwing away their fortunes in gifts to a debased urban mob. They were commonly men of exceptional distinction who could be as heroically selfless in political causes as their ancestors were heroically valorous in military ones. As Werner Jaeger was to point out, Athens did not destroy its aristocracy but rather tried to turn its entire citizen body into one.

This pithy formulation has several levels of meaning. Politically, it sums up the slow but sweeping way in which Attica's "bribe-devouring" nobles were shorn of their power. By the seventh century B.C., Athens and its environs were on the brink of revolution. Plutarch tells us that the "common people were weighed down with debts they owed to a few rich men. They either cultivated their lands for them and paid them a sixth of the produce and were hence called *Hectemorioi* and *Thetes*, or else they pledged their own persons to raise money and could be seized by their creditors, some of them being enslaved at home, and others being sold to foreigners abroad. Many parents were even forced to sell their own children (for there was no law to prevent this) or to go into exile because of the harshness of their creditors. However, the majority, which included the men of most spirit, began to make common cause together and encourage one another not to resign themselves to these injustices, but to choose a man they could trust to lead them. Having done this, they proposed to set all enslaved debtors free, redistribute the land and make a complete reform of the constitution."[5]

At this point, Athens might have easily gone the way of Rome. Five centuries later; the Gracchi brothers, who faced a nearly identical crisis between bitterly polarized classes of landless peasants and bloated patricians, raised such sweeping demands for political change and agrarian reform that they opened irreparable wounds in the Roman body politic. It remains to the credit of the Athenians that the crisis was handled very gingerly. The moderation that Hellenic society turned into a deeply personal as well as civic ethos was to find its embodiment in Solon, a noble of considerable prestige who had earned the respect of the Athenians as a whole. No one could have contained the crisis that faced his people with greater prudence.

Elected archon of the *polis*—its chief magistrate—and invested with sweeping powers to resolve the conflict, Solon followed the middle course that eluded the Gracchi and the more sincere *populares* in the closing years of the Republic. For the poorest of the *demos*—the *Hectemorioi* and *Thetes*—he removed their most pressing economic burden by canceling all outstanding debts and making debt slavery illegal. To strengthen their political status, he revived and expanded the functions of the *ekklesia*, which had virtually ceased to exist since tribal days. The assembly was authorized not only to enact the community's laws and elect its magistrates; it convened as a court of justice to deal with all cases other than homicide, a crucial advance in empowering the *demos*. The upper crust of the nobility—the *Eupatridai*—were obliged to relinquish their hereditary claim to furnish Athens with its archons, a powerful, annually elected magistracy whose number Solon increased to nine. The office, to be sure, could only be filled by landowners, but the door to executive power, which later generations were to open, was now unlocked for the *demos*.

Solon never pretended that he desired the political and economic supremacy of *demos*, nor did he try to divest the nobility of power. As his verses indicated, he shrewdly steered a middle course through a crisis that would have exploded in social chaos had any of the contending orders gained absolute supremacy over the others.

> *To the mass of people I gave the power they needed,*
> *Neither degrading them, nor giving them too much rein;*
> *For those who already possessed great power and wealth*
> *I saw to it that their interests were not harmed.*
> *I stood guard with a broad shield before both parties*
> *And prevented either from triumphing unjustly.*[6]

The land magnates were not deprived of their holdings, nor were the *Hectemorioi* permitted to reclaim the sixth of their produce that was taken by their landlords. Usury still plagued the Attic peasantry, although the lender and borrower were more evenly squared off by the controls that the *ekklesia* exercised over legal disputes. Solon also created a Council of Four Hundred—the famous Athenian *boule*—to which only propertied men could be elected annually. However, the *boule* served to check not only the popular *ekklesia* by rigorously determining its agenda and supervising its deliberations; it also checked the behavior of the aristocratic Council of the Areopagus (formerly the powerful Council of the Eupatridai), whose functions over the years were to become more ceremonial than political. A number of reforms, unique for their time, were made to expand individual rights and alter popular etiquette. An heiress, burdened by an impotent husband, was free to marry his next of kin, and relatively poor women were spared the need to collect dowries. Individuals could will their property as they chose, not according to familial dictates—a law that struck an important blow at the collective solidity of the aristocratic *genos* and its concentration of sizable wealth. A number of lesser laws restricted displays of excessive luxury and riches. But perhaps the most strikingly Hellenic law imputed to Solon was one that disenfranchised any citizen who, to use Plutarch's words, "in the event of revolution, does not take one side or the other." It was Solon's intention, Plutarch emphasized, not to reward citizens who seek the safety of neutrality, apathy, or indifference at the expense of the public interest. Athenians were expected to be politically involved, irrespective of the causes to which they adhered, or else they were not worthy of citizenship. It was an affront to the Hellenic concept of citizenship for a man to prudently and selfishly "sit back in safety waiting to see which side would win."

Characteristically, Solon refused to linger on as a tyrant—by no means a pejorative term or a despised status in those days—despite pleas that he become one, and went into voluntary exile for ten years. His work did not resolve all the political and economic problems that brought him to the archonship, nor did it prevent a tyrant from ultimately replacing him. But it provided Athens with the opportunity to absorb his reforms and the time needed for its citizenry to mature politically—to use and accustom itself to the *ekklesia*, to learn the arts of compromise, and to develop a political etiquette that fostered a sense of civic commonality rather than social conflict. The rather mild tyranny of Peisistratus and his son Hippias, which followed Solon's departure, greatly reduced the power of the Attic nobility and initiated the economic and political changes that were to lead to the democracy. Recalcitrant nobles were forced into exile and their estates divided among the poor. The needy were given livestock and seed, exports were promoted, and a vigorous foreign policy opened new markets to Athenian commerce. Peisistratus, despite the personal control he exercised over Attica's affairs and his blatant nepotism, adhered to the Solonic constitution so meticulously that even Solon was induced to support him after his return to Athens. The political level of the Athenian citizen was raised enormously by the tyranny, and the commoners—farmers, shepherds, artisans, and merchants—benefited so considerably from the archonships of Peisistratus and Hippias that the tyranny became self-vitiating. Three generations had passed since Solon had been given the archonship, and Athens was now ready for a democracy. In terms of the mere flow of events, the efforts to initiate authentic popular rule seemed to come almost as a reaction to attempts by the aristocracy to restore their old clannish oligarchy, and many of the democracy's features must be seen as institutionalized efforts to prevent the emergence of entrenched power by elites of any kind. But the Athenian people, too, seem to have become more certain of themselves and their capacity to govern their own affairs. The aristocracy, in turn, appears to have suffered a genuine lack of nerve. After a feeble attempt to eliminate the reforms of Solon and the Peisistradae, the nobles broke ranks. One of their own kind, Kleisthenes, the head of the aristocratic Alcmaenodae, became archon in 506 B.C., and the democratization of Athens was launched in earnest.

Kleisthenes struck decisively at the societal basis of aristocratic power—the traditional kinship network that gave the Attic nobility its very sense of identity. The ancient Greek phratries and clans were simply divested of any political power and gradually declined in importance for want of any significant functions. The old Ionian system of four ancestral tribes was converted into ten strictly territorial "tribes" based exclusively on residence. The villages and towns of Attica, in turn, became outlying sections of Athens and were designated as "demes" instead of *genoi*. Politics now became inseparable from territorialism: the *demes*, with their own popular assemblies, were grouped together in varying numbers into thirds or *trittyes*, and three *trittyes*, in turn, constituted a tribe, hence Attica was composed of thirty *trittyes*. Ten of the thirty *demoi* were composed of residents in or around Athens; another ten, from the maritime districts; and the remainder from the interior.

Kleisthenes shrewdly placed one urban *trittys* in each of the ten tribes, so that the Attic agrarians from whom the nobles garnered whatever popular support they had were politically buffered by city citizens—the men who were to form the backbone of the democracy. This switch in the governance system of the *polis* was strategic: it fostered the power of a citizenry that was distinctly urbane, cosmopolitan, and forward looking, vitiating the strongly hierarchical structure of a once-entrenched, highly parochial, feudal class system. At the same time, tradition was kept alive by using the language of the tribal world (even the word *gene* had a special clannish origin), retaining a number of local religious associations, chieftainlike figures such as *demarchs* (the deme's version of the Athenian archons), and by making membership in a deme hereditary even though a citizen might choose to reside at some later time in another part of Attica. Kleisthenes, in effect, "revolutionized" Athenian political life in the literal sense of the term: he replaced a once-egalitarian tribal system that had been perverted into a harsh feudal hierarchy with a tribalistic structure that actually restored the old freedoms of the people on an entirely new political and societal level. Athens had revolved in a full circle—more precisely, a spiral—to the *isonomia* of its tribal past, but without the innocence that made the early Greeks vulnerable to hierarchy and domination.

The *boule* was increased from a council of four hundred to five hundred and restructured so that fifty men from each of the ten tribes rotated every tenth of the year as an administrative "executive committee" between sessions of the *ekklesia*. Each tribe selected its fifty *bouleutes* by lot, a practice that became so widespread that even archons were so chosen from members of the *boule*, as were members of Athenian juries (*dikastoi*) and lesser functionaries. Apart from the *polis*'s magistrates, no property qualifications debarred Athenian citizens from participating in the governance system, and under Perikles the last restrictions that lingered on from Kleisthenes's reforms disappeared completely. In time, members of the *boule*, the *ekklesia*, and the *heliaea* or courts were compensated for participating in these institutions, generally on a *per diem* basis and in the case of the *boule*, annually. No public office could be held for more than a year, and with certain exceptions (jurymen and generals) none could be held more than twice in a lifetime.

This extraordinary opening of public life to the Athenian citizenry was completed during the sixty years that saw Kleisthenes assume the archonship in 492 B.C. and the outbreak of the Peloponnesian War in 431. An analytic account of the democracy, its possibilities and its limitations, properly belongs to the discussion of civic freedom that follows in the next chapter. For the present, what counts is the way the democracy formed the citizenry and, in turn, was formed by it. Democratic leaders such as Kleisthenes and Perikles did not foist a remarkably open system of participatory governance on a passive people; the institutional structure and the body politic interacted closely with each other against a haunting background of quasitribal social forms and relationships. This is politics at its best—in a lived sense, not a formal one. The Athenian notion of *arete*, the daily practice of *paideia*, and the institutional structure of the *polis* were synthesized into an ideal of citizenship that the individual tried to realize as a form of self-expression, not an obligatory burden of self-denial. Citizenship became an ethos, a creative art, indeed, a civic cult rather than a demanding body of duties and a palliative body of rights. At his best, the Athenian citizen tried not only to participate as fully as possible in a far-reaching network of institutions that elicited his presence as an active being; the democracy turned his participation into a drama that found visible and emotional expression in rituals, games, artwork, a civic religion—in short, a collective sense of feeling and solidarity that underpinned a collective sense of responsibility and duty. This drama extended beyond life itself. The Athenian citizen had little hope of any certain immortality other than the memory he left behind in the *polis*. Afterlife became a form of political life and eternality existed only insofar as noble political actions were memorable enough to become part of the *polis*'s history and destiny.

The *polis* shrewdly availed itself of aristocratic attributes to bring the individual Athenian into the full light of citizenship, with its high standards of civic responsibility. We have seen how the *agora* prepared the way for the *ekklesia*. By the same token, the gymnasium, presided over by a *paidotribes,* extended the all-consuming fascination of the Greeks for athletics into martial arts, thereby preparing the young for military training. Troubling as this may be in modern eyes, war was a fact of life in antiquity. Apart from certain cults and religions, pacifism found no following among the ancients. One either fought unquestioningly for one's city or faced clientage and slavery in the event of defeat. The citizen-soldier was much more than a pillar of Athenian military strategy. He was a guardian not only of the *polis* but also of the democracy, as we shall see later, just as the citizen-farmer became the embodiment of its ideal of *autarkeia.*

But the *polis,* particularly the democracy, gave the family's aristocratic attributes a uniquely public character. And it is in this non-elitist sense that the democracy elevated its citizens into an "aristocracy"—not only as a result of the participatory nature of its courts, assemblies, and councils, but also the familial mood of *koinonia* that its civic festivals engendered. Without losing their essentially religious character, rituals and festivals became a form of politics. The *ekklesia* opened with prayers, and its agenda was composed of sacred as well as secular topics. Problems of constructing temples or planning festivals occupied the assembly as earnestly as matters such as ostracism or the ratification of treaties. In famous funeral oration, which so vividly captures the spirit of the democracy, Perikles cites among its attributes "the contests and sacrifices throughout the year, which provide us with more relaxation from work than exists in any other city."[7] Webster's observation that the Athenians tended to turn their holy days into holidays suffers from a certain degree of misunderstanding.[8] Athenian life struck a remarkable balance between religiosity and secularity: the camaraderie of participating in a spectacle or sharing fully in the excitement of the games imparted a quasireligious sense of communion to civic life. This sense of communion more accurately conveys the intense meaning of *koinonia* than do such commonplace words as "community." At the risk of repetition, we could say that such shared experiences unravel the Athenian notion of "community" as an on-going activity of communizing, not simply an institutionalized form of participation. Like the *agora,* the "contests and games" of Athens often created a shared sense of preternatural civic enchantment.

The democrats knew this very well and these "contests and games" occurred with considerable frequency. To discuss them in detail would require a volume in itself. Let it suffice to say that every year, many days were devoted to the Lesser Mysteries in February and the Greater Mysteries in September, rituals that centered around Persephone's descent into Hades and Demeter's mourning, the mythical explanation for the occurrence of winter and its lean months. Every July, Athenians participated in the Lesser Panathenaia, which culminated quadrennially in the Greater Panathenaia, an extraordinary parade of Athenians and Athenian life in full array, if we are to judge from the bas relief that girdles the Parthenon. Almost every month, Athenians witnessed or participated in a variety of rituals, contests (athletic, musical, poetic, and choral), or celebrations to honor deities, historic events, great personages, victors in Panhellenic festivals such as the Olympics or the fallen of past and recent battles. Religion and civic loyalty blended the great variety of personal and social interests within the body politic into an underlying commonality of outlook that, if it did not remove serious conflicts, rarely reached such desperate levels that they could efface the democracy from within. Ultimately, it was to be Macedonian rule that brought the democracy to its definitive end, not the Athenians. For all its shortcomings, the democracy in various forms persisted through nearly two of the most stormy centuries of the ancient world and, at its height, exhibited a degree of cultural and intellectual creativity that has no peer in western history.

Perhaps the most important of the Athenian festivals was a comparatively new one: the City Dionysia. Even more than the Greater Panathenaia, when all of Athens went on display with a large tapestry (the *pelops*) that depicted the triumph of Olympian "reason" over the chthonic rule of "force," the City Dionysia was strongly democratic in its focus. It was then, for three out of a span of six days that overlapped March and April, that Athenians could witness the great dramatic tragedies that gave the democracy its ideological meaning. By the thousands, Athenians flocked to the Theater of Dionysos on the southeast slope of the Acropolis to see the plays of Aeskylos, Sophokles, Euripides, and others who literally created serious drama in western culture. Under the clear skies of a high Mediterranean spring, they watched with absorption the Aeskylean drama of their own *polis* and its human antecedents unfold with a majesty that may have verged on reverence—certainly a thrilling sense of exaltation that compels a modern historian of the democracy, W. G. Forrest, to cite Aeskylos "as the greatest of the three Athenian tragedians," and for many, including myself, the greatest of the world's dramatists.[9] Aeskylos's trilogy, the *Oresteia*, advanced a powerful validation of the democracy that, in its emotional and declaratory power, may even exceed the funeral oration of Perikles—a trilogy that was constantly replayed and kept winning prizes at the City Dionysos long after the author's death.

Its story has been told and interpreted repeatedly, and to explore it at length would be a tiresome redundancy. Let it suffice to say that the murder of Agamemnon, the returning chief of the besiegers of Troy, by his wife, Klytemnestra, followed by her own death at the hands of her vengeful son, Orestes, opens the whole drama of Athens's transformation from a quasitribal society, rooted in kinship rules, custom, and chthonic deities, into a political community—a *polis*—based on residence, reason, and the anthropomorphic Olympians. It is Athene who, in a challenging statement against the Erinyes (the three female guardians of "matriarchal" blood ties and tribal retribution for the murder of one's kin), solemnly declares:

> It is my task to render final judgement here.
> This is a ballot for Orestes I shall cast.
> There is no mother anywhere who gave me birth,
> and, but for marriage, I am always for the male
> with all my heart, and strongly on my father's side.
> So, in a case where the wife has killed her husband, lord of the
> house, her death shall not mean most to me. And if the other votes
> are even, then Orestes wins.[10]

After being pursued by the Erinyes for committing a blood

crime more damning than a marital one—and particularly against his mother from whom early tribal descent may have been traced—Orestes is absolved and the Erinyes reconciled by acquiring a civic status as Eumenides, the kindly ones who look after the well-being of the Athenian *polis*.

The trilogy has many levels of meaning, probably all of which had a gripping effect on the audience that knew Aeskylos personally or, in later years, by reputation. The *Eumenides*, the last drama of the trilogy, celebrates the victory of civic law and rationality over tribal custom and unthinking mimesis. Athene, born of Zeus's head, embodies *logos* and justice. In a strongly patriarchal society that saw male rationality as the sole bulwark to dark chaos and an uncertain, untamed world, it was not difficult to identify "fickle" woman with nature and the *polis* as the sole realm of freedom and law. Orestes's trial, which marks the culmination of the trilogy, is presented as a new dispensation in the affairs of men. The Erinyes unrelentingly pursue any homicidal perpetrator of the blood tie. All that counts with them is the act of murder of one's kin, not the circumstances or motive that produced it. To the Athenians, this behavior was evidence of unreasoning tribalism, of an irrationality that precluded any civic union based on discourse, logic, and orderly compromise. Recourse to trial rather than ordeal, to the weighing of circumstantial evidence, motive, and logical judgement rather than mere action and contests of fortitude, marked the first step toward a political *koinonia*—the city as a *polis*.

To the Athenian audience, which believed Athens was the first city to establish a system of laws, the trial of Orestes was a founding act. Athens's decision to try homicide by a judicial process literally created the *polis* as an ethical union based on justice. "If it please you, men of Attica," intoned Athene, the patroness of the *polis*, "hear my decree now, on this first case of bloodletting I have judged. For Aegeus' population, this forevermore shall be the ground [the Hill of Ares] where justices deliberate.... No anarchy, no rule of a single master.... I establish this tribunal. It shall be untouched by moneymaking, grave but quick to wrath, watchful to protect those who sleep, a sentry on the land."[11] Henceforth, justice, not tradition; reason, not custom; fact, not ordeal; motive, not myth are to guide the men of Attica, for without this new dispensation, the *polis* has no ethical meaning, nor does the democracy have an ethical rationale.

It is easy to see in Aeskylos not only the clearest voice of the *polis*, of a body politic free of arbitrary rule, but also of the democracy—whether as a "radical," as Forrest declares, or as a "revolutionary poet," as George Thomson seems to believe.[12] To the Athenians, who apparently revered Aeskylos, the *Oresteia* unfolds the emergence of justice from a hazy "dark" world of tribal antiquity and its fortunes in the arbitrary domain of warriors, nobles, and the *genoi*, into the clear light of the *polis* and its orderly citizenry. The identification of the audience with the drama must have been intensely personal; it was the authentic protagonist of the play. Of Aeskylos's remaining dramas, *Prometheus Bound* arrests us to this very day with its message of heroic defiance against unfeeling authority and its expansive belief in humanity's sense of promise, indeed its capacity to advance toward an ever-wider horizon of intellect and wisdom.

It would be easy enough to end our discussion of the Athenian drama with Aeskylos, but Sophokles beckons us with other facets of meaning that must have deeply affected the Athenian spirit. His *Antigone* raises the ambiguities of justice in the form of conflicting individuals, indeed personalities. Antigone herself becomes the embodiment of tribal kinship rules in her frenzied zeal to bury the body of her brother, Polyneikes, whom Kreon, the King of Thebes, wishes to leave to vultures and dogs as an example for future rebels. Kreon thus appears as the embodiment of secular authority, the civil counterpart to Antigone. The drama emerges as a fugue, in which deeply emotive ethical themes are played against each other and intermesh. If Aeskylos's characters tend to appear as forces rather than individuals, despite Klytemnestra's awesome personality and Athene's majesty, Antigone wins us as a persecuted heroine who seeks the burial of her brother's body and Kreon as the willful embodiment of civil rule, prior even to the rise of the *polis*. For we are still in the time of the royal *oikos* or household. In writing the drama, Sophokles may have tried to show that after the decline of this prepolitical world of blood clans and regal palaces the Athenian citizen no longer had to confront the pangs of tragedy in Hegel's sense of the term—a drama in which both sides are right. The *polis* spares us the need to pit divine law in the form of tribal commandment against human law in the form of civil retribution for rebellion. We can read our romantic sympathies for Antigone into the drama, but a Greek audience might have viewed her differently. It would have been obliged to place Kreon's civil obligation on a par with Antigone's tribal obligation, and its sympathies would have been allotted according to political as well as personal sensibilities. Relief emerges from the sense of delivery that a concealed theme in the drama affords: the free air of the *polis* in which the citizen can breathe without the presence of such exotic conflicts.

Such is also the case in one of Sophokles's most celebrated—and possibly one of his most misunderstood—plays, *Oedipus the King* and *Oedipus at Colonus*. Here man, potentially the citizen, rises to a clear and level gaze at his fate and his ability to learn from suffering. If there is anything that impresses us about the play, it is the fact that Oedipus emerges with greater nobility toward the end of his life, all his misfortunes notwithstanding, than in its portentous beginnings. No broken King Lear confronts us but rather a hero whose passion for truth ultimately transcends the disastrous impact of patricide and incest. The cathartic core of the drama is unmistakable: it is the nobility of the individual, purged of archaic curses and trammels, of man's high promise and destiny to achieve insight and wisdom when he acquires the status of a free citizen.

Yet, the Greek ideal of the citizen, in contrast to ours, is not monadic. However individuated the Greek drama became with Sophokles and even more decidedly so with Euripides, the Athenian citizen would have mocked the entrenched bourgeois myth that the free man is an atomized buyer and seller whose choices are constrained by his own psychological and physical infirmities. He would have seen beyond the arrogance of this self-deception into the pathos of the bourgeois citizen's clientage to the powerful, his aimless pursuit of wealth, his reduction of life to the acquisition of things—in short, a *moira* or destiny governed more by *ananke* or necessity than by the understanding that even such a nascent personality as Oedipus

brings to his own insight into reality. Such a despised bourgeois being, he would have concluded, is no less archaic than the Erinyes,
who must be freed from their primality as mere forces and recon-
structed by reason and justice into the "kindly ones." Only then, in
synchronicity and ethical union with the *polis*, is citizenship more
than a formality and its practice more than a ritual.

By the same token, the Athenian citizen was not a corporate being in our usual meaning of the term. Most present-day discussions
of the Athenian's lack of individuality and the *polis*'s tendency to
subserve his personality to an overbearing collectivity are weighed
down by Eurocentrically neurotic images of the individual as such.
The modern identification of individuality with egotism and personality with neurosis has been overindulged under the rubric of
"modernity" with an arrogance that bears comparison only with the
conceited claims of psychoanalysis and psychohistory to explain
the human drama in all its aspects. That human beings can be individuated in different ways—some as highly social and political
beings, others as private and self-indulgent beings, still others as
combinations and permutations of both—is as alien to the imperialistic claims of "modernity" as it is to the admirers of *gemeinschaft*,
the stagnant folk community based on kinship and organismic relationships. To the Hellenic citizen of a *polis*, leaving all its mythic
origins aside, the monad would have seemed as prehuman as the folk
community seemed prepolitical. Individuality meant citizenship.
And, ideally, citizenship meant the personal wholeness that came
from deep roots in tradition, a complexity of social bonds, richly
articulated civic relationships, shared festivals, *philia*, freedom from
clientage and freedom for collective self-determination through institutions that fostered the full participation and everyday practice
of a creative body politic. To be such a citizen, one had to live in a *polis*—a city that possessed an *agora*, a space to convene general assemblies of the people, a theater to dramatize the reality and ideology of
freedom, and the ceremonial squares, avenues, and temples that gave
it reverential meaning. To remove any of these elements that made
up this whole was to instantly destroy it. Without every one of them,
cultivated on a daily basis by the *paideia* of citizenship and guided

by an unerring concept of *arete*, the Athenian ideal of citizenship fell apart and its institutions became hollow forms.

But such notions, valuable as they may be, lack a processual content that sees this whole from the standpoint of an ever-fuller development. We can see this by looking closely at the quixotic nature of the Greek dramatic tragedy itself—the staged experience of the *polis* and citizen. For it was through the tragedy that the Athenian underwent self-examination, much as though he could withdraw from his own skin and observe himself thoughtfully for what he really was. It is notable that all the tragedies we have at our disposal are made up of mythic and epic material. As civic experiences, they seem to function dissonantly with stuff that is drawn from a prepolitical world, from eras that precede the *polis* in its most advanced stage of development. Their heroes and heroines are Bronze Age characters who still live under the commandments of a tribal or warrior society, by canons of right and wrong that are archaic at best and chthonic at worst. In the transformations that the dramas recount from one world to another, we rarely seem to rise from new foundations embedded in still older ones to the edifice that the audience knew at hand—the *polis* of the fifth century B.C. and its democracy.

Yet the antiquity of the characters, material, and themes with which all the great tragedians seem to have worked is belied by a startling sense of innovation, by tensions from which new ideals, institutions, and credos are born. The *Oresteia* seems to not only justify radically new communal dispensations, indeed most of which already existed when the trilogy was presented; its radicalism consists in the fact that Athene's decrees are fraught with the promise of change, with motives that are likely to yield ever-wider ethical and political horizons. One senses that a community guided by justice has just begun to emerge, that the civic ideal of freedom is still nascent, that citizenship and the *polis* are still experiencing birth pangs and their future still lies before them. Few Athenians, I suspect, could have left such dramas without feeling an intense sense of futurity and a serene commitment to growth. *Prometheus Bound* is even more challenging in its thirst for heroic innovation and its passion for a fuller and richer experiential life, as are the Oedipus dramas for truth at all costs.

Hence, seemingly conservative material is played, replayed, and ultimately transmuted into searingly progressive hopes and possibilities. When we bear witness to these dramas by viewing them in all their starkness as idealized "archetypes" of an immutable human nature, we do violence to the tensions they probably produced in the Athenian audiences of their day. These plays were no mere spectacles or cathartic outlets for pent-up civic anxieties. Quite to the contrary: they were sources of motivation and tension, designed to move their audiences forward to noble deeds and great enterprises, to contrast past with present so that meaning, continuity, and tendency could be imparted to the future. They voiced a hope in the human spirit that belies the conventional interpretation of the Athenian drama as fatalistic in its view of life and resigned in its image of destiny. Indeed, it is with Euripides, that most "modern," "realistic," and "individualistic" of the Greek tragedians, that hopelessness, alienation, and the insufferable burden of circumstances seem to devour the characters of the drama and exhaust the hopes of the audience. Here, the *deus ex machina* snaps up the immanent dialectic of the older drama with its promise of futurity and hope.

The *polis* was no less a theater for the practice of virtue than the orchestra at the foot of the Acropolis was a home for the performance of the plays it watched. And the citizen was no less an actor in a great civic drama than the men who performed for him with masks in the City Dionysos. In both cases, the plot we call the history of Athens incorporated the layered traditions that formed its cultural biography and ideology. So central was the citizen to this plot with his "freedom for" as well as his "freedom from," that the histories of the time, when they refer to Attica, speak more commonly of "the Athenians" than they do of "Athens." Unerringly, they reveal that Athens was no mere collection of structures, no simple geographic locale, that any aggregate of people could occupy without the *polis* losing its authenticity. Admittedly, the city outlived the *polis*, and the democracy in a formal sense outlived the citizens. Democratic institutions persisted in a truncated form long after Athens's final defeat by the Macedonians at Krannon in 322 B.C. No city was a *polis*, in Aristotle's view, unless it had an *agora*; but, needless, to say, no *agora* could have produced a democracy unless it had the kind of citizen the historians of that day called "the Athenians."

There is a myth that the classical Athenian *polis* was the spontaneous product of custom and tradition. Hence, its integrity rested largely on hidden presuppositions of which it was largely unconscious, indeed, to which it was sublimely oblivious. This notion of an unreflective Athenian morality, imputed to Hegel and held by many romantics in the last century, is belied by almost everything we know about the democracy. The Aeskylean tragedies and Perikles's funeral oration, to cite two highly important and widely separated examples, completely refute the image of the democracy as a phenomenon based on mere habit and convention. One must totally ignore Kleisthenes's reforms and the subtle melding of religious tradition with clearly formulated political goals not to recognize that the democracy was a consciously crafted structure, the product of purposeful, insightful, and thoughtful efforts to achieve clearly perceived goals—and, let me add, guided by a philosophy and practice of its own. The introduction of the lot, the rotation and limitation of tenure in public office, the development of the assembly and the *boule*—all reveal a degree of intentionality and clarity of purpose that has few equals in any later constitutional developments. That Sokrates was executed for questioning this order on charges that were archaic even by conservative standards of fifth-century Athens has more to do with the politics of the trial, particularly Sokrates's arrogance and his desire for martyrdom, than the challenge his philosophy posed to the *polis*'s "ethical order." No political community was more aware of its uniqueness and wondrous of its achievement than the Athenian democracy. And no one was more consciously committed to this humanly wrought body of institutions than the Athenian citizen.

By contrast, later ideals of citizenship, even insofar as they were modeled on the Athenian, seem more unfinished and immature than the original—hence, the very considerable discussion I have given to the Athenian citizen and his context. As we shall see shortly, there were impressive attempts to create patterns of civic freedom that approximated the democratic *polis* in medieval city-states and in the American and French revolutions. These attempts were usually intuitive. The Americans who attended town meetings in New England after 1760 and the French *sans-culottes* who filled the radical sections of Paris in 1793–94 were hardly aware of the political drama that unfolded in Athens some two millennia earlier, although their more knowledgeable spokesmen fingered Plutarch throughout their lives and occasionally adopted heroic names during fervent moments of the two revolutions. What was lacking in their attempts to achieve a living approximation of civic democracy was not merely ideology. Greece, commonly Athens (although Sparta's garrison-state was a close runner-up for some utopian writers and for Rousseau), was held up as a fascinating experiment in popular self-governance and, as usual, compromised with the republican ideals generated by the emerging nation-state.

For this reason, the Athenian ideal of citizenship, like its parallel creation, politics, demands careful exploration today even more than at any time in the past. If moderns find democratic politics and citizenship a desideratum, they will never achieve them without a supreme act of consciousness. They must not only want it but know it. Athenian civic goals, for all their shortcomings (notably Athens's treatment of women, alien residents, and its widespread use of slave labor), must be rooted in an everyday notion of what we mean by politics. Is it statecraft? Or does it center around social entities such as cooperative, vocational societies or tribes in the countercultural sense of this much-abused term? Or some broad concept of grassroots organization that passes under words such as "localism," "decentralism," and "bioregionalism"? Is it an educational activity—a civic *paideia*—that fosters the citizen's empowerment, both spiritually as well as institutionally? Or is it primarily a form of "management" whose goal is administrative efficiency and fiscal shrewdness?

No modern body of ideas, to my knowledge, has wrestled with the answers to these questions adequately enough to draw clear distinctions among the social, the political, and the statist so that a meaningful outlook can be formulated—one that will seek the delicate balance of ingredients (traditional, familial, ethical, and institutional) and the *paideia* that articulates an authentically democratic politics with a concept of citizenship that gives this outlook reality. Nor do we have a clear idea of the extent to which the city, properly conceived as a humanly scaled ethical community, differs from urbanization and the inhuman scale produced by the nation-state. We rarely understand how integrally an ethical politics is wedded to a comprehensible civic scale, to the city itself, conceived as a thoroughly manageable and participatory union of citizens, richly articulated by tradition and by social, cultural, and political forms. We live so contemporaneously within the given state of affairs, the overbearing "now" that eternalizes the status quo, that no society is more prey to the workings of mindless forces than our own. Bereft of a serious regard for history, indeed for the experiences of our own century, we find ourselves in the airless vacuum of an immutable "present," a time warp that precludes any sense of futurity and ability to reason innovatively.

Accordingly, where thinking of a crude kind does exist, lines of thought fecklessly crisscross each other like the scrawlings of an infant that can barely grasp a pencil. Oligarchies are accepted as democracies; virtual monarchies as republics; state institutions as social forms; social forms as political ones. One can hardly speak, here, of a kind of shallow eclecticism that afflicts every age; rather one encounters a potpourri of unfinished notions, barely worthy of being called "ideas," that are mingled together in the viscera rather than the head. To think out ideas to their logical conclusion, to be consistent and complete in thought, is viewed as a form of intellectual aggression that stimulates reticence or withdrawal rather than dialogue or response. The modern "ego" is now so fragile that the mere presence of passion, particularly in the form of argumentation, is a stimulus to flight or, worse, cause for a yielding quietism. The emphasis that the Athenians placed on speaking, which was identified with democracy, spells out the extent to which a qualitative breach separates our pitiful image of public activity from the classical one of two millennia ago.

What should be stressed, for the present, is that the Athenian ideal of citizenship did not die in the formal sense, even after Macedonia had eviscerated the democracy. The *ekklesia* and *boule* functioned for generations after Athens lost her independence; the institutions were kept alive as a political sop to the Mediterranean-wide belief in Athens's uniqueness, all the more to cloak a growing despotism that finally caused the last institutional vestiges of the democracy to wither completely. And here lies a lesson that later institutionally oriented democrats were to learn at great cost. An *agora* does not in itself produce a *polis*, nor does an *ekklesia* in itself produce citizens. All of them must exist, to be sure, if a democracy is to be established; but they remain mere formalities if they fail to interact in the proper way to form a unified whole. It was this largely formal sense of democracy and citizenship that served, in part at least, to abort the only attempt to rescue the Roman republic in the guise of the Athenian *polis*. It would be difficult to tell from Plutarch's biographies of Tiberius and Gaius Gracchus that the brothers were no mere agrarian reformers; indeed that their laws to supplant the Senate's powers by the authority of the *comitia tributa* came as close to a fundamental political revolution as Rome ever experienced. The brothers, in M. I. Rostovtzeff's view, were clearly out to "set up at Rome a democracy of the Greek type," a "dream or farce," as he calls it, that was designed "to transfer from the Senate to the popular assembly the decision of all important business, or, in other words, to set up at Rome a democracy after the Athenian model."[13] Nor was this goal entirely impractical, at least institutionally. "To secure this point no special law was required: according to the constitution all important business was, in theory, settled by the popular assembly; the innovation was this, that business which by custom had hitherto been decided by the Senate was now laid by [Gaius] Gracchus, as tribune, before the popular assembly, for their consideration and decision."[14]

The effort ultimately foundered not for want of institutional mechanisms but for lack of a genuine citizenry. Despite Rostovtzeff's aristocratic biases, he is on solid ground when he emphasizes that "to allow land even to every member of the proletariat could never bring back a time when the state rested securely upon a strong peasant population."[15] The culture that produced men such as Cincinnatus or the elder Cato was waning and had to be recovered or a new one developed in its place if mere institutional changes were to be viable and lasting. The Gracchi would have had to deal with still another change that would ultimately undermine the entire republican edifice. Rome was no longer a city in any Hellenic sense of the term. It was the center of an emerging empire, an *urbs*, to use the conventional Latin word for "city," that had ceased to be a *civitas*, the word Romans commonly used to denote a "union of citizens." The terminological distinction is apt: the Roman *urbs* was growing uncontrollably by the second century B.C., at the expense of the Roman *civitas*; it was sprawling outwardly in size as well as worldwide in scope, especially after the spoils of the Punic Wars began to fill its coffers. Conceivably, the yeoman-citizens who founded the republic could have turned it into a democracy. But once they "came down from the Seven Hills" on which Rome was founded, they became "small," to use Heine's words. The "idea of Rome" as a spiritual heritage diminished in direct proportion to the growth of the city. "The greater Rome grew," Heine wrote, "the more this idea dilated; the individual lost himself in it: the great men who remain eminent are borne up by this idea, and it makes the littleness of the little men even more pronounced."[16]

The decline of the ancient world did not efface the ideals that Athens and Rome inspired. Later eras were to blur them so that we now speak of two kinds of "democracy," "direct" and "representative," as though they derive from a common heritage. During inspired moments, medieval and modern cities were to fluctuate between the two and, at times, come closer to a Hellenic democracy and citizenry than the Gracchi could have achieved in Rome. What these cities achieved in fact, however, they rarely achieved in ideology; hence their patterns of civic freedom were basically intuitive creations. They did not last long. But in an era of chilling discontinuities, when our knowledge of western culture is in grave peril of disappearing, we can no more ignore the legacy of their example than we can ignore the sprawling urbanization that threatens our civic identity and political freedom.

Chapter Five
Patterns of Civic Freedom

Although the city was the cultural center of the ancient world, its material base was extremely fragile. Hence the remarkable rise and fall of cities throughout antiquity. City life depended to a remarkable degree on the viability and the economy of agrarian communities and labor so that cities appear more as islands in a vast rural ocean than self-sustaining economic entities. Indeed, prudent estimates of the number of rural workers needed to sustain a single city dweller have been placed at a ratio of ten to one. Hence any serious agricultural catastrophe like a drought, floods, or pest infestation, not to speak of wars that devastated agriculture, could lead to the breakdown, indeed the complete disappearance, of a major city as so many civic ruins from antiquity attest to this very day.

The fragile dependence of the city on the countryside's economy placed a high premium on the stability of urban institutions. The Hellenic emphasis on the *polis* as a realm of reason that had to vigilantly guard itself against a surrounding world of chaotic uncertainty and mindless barbarism—a sensibility so important to understanding its attitude toward women, slaves, and aliens—is not without its roots in the compelling realities of ancient urban life. A nature tamed by man, specifically by well-established and reliable farmers, forms the matrix of Hellenic philosophy and the great cultural monuments of the Athenian *polis*. Civic democracy was largely an agrarian democracy, and its ideal of the farmer-citizen, however dubious it became in the later history of the *polis*, was the product of a very viable agrarian heritage. Rome, too, developed these ideals and did not require Greek culture to produce them.

Correspondingly, the ancient city was not only a center for country life whose standards were formed by farmers, later by nobles and "country gentlemen," but it was "seafaring" in a very loose sense of the term. Owing to the insecurity that the city felt as an "atoll ... on an ocean of rural primitivism," it direly needed access to the sea to cope with any dislocations between town and country. The all-important factor that could correct any imbalances caused by drought, flood, plague, and other such disasters was imports or, put more bluntly, the tribute that a city-state could exact from a league of captive "allies" or the victims of outright conquest. The Delian League, which Athens established to assert its supremacy over the eastern Mediterranean, quickly turned into a relatively modest looting enterprise to feed and provide the "good life" for its citizens. The Roman Empire is a lasting historical example of parasitism run riot to the point of moral, cultural, and societal suicide.

Hence, ancient cities were located within a hundred miles or so of the Mediterranean Sea or close to navigable rivers that led to major bodies of water. Ancient Lugdunum (modern Lyons) provides us with an example of a city that Rome founded in 43 B.C., deeply inland in Gaul, that was felicitously located at the confluence of the Rhone and Saône rivers. Beyond the coasts and rivers, cities dwindled to towns and, not much farther, to villages and hamlets that then phased into a little known wilderness where military forts kept a guarded eye on restless barbarian tribes. Ancient cities were never autonomous urban centers. Fragile in the extreme, they depended upon local agricultural communities for the most important essentials of life, a very simple technology, and a considerable amount of human muscle power to work it. If they expanded beyond the capacity of their agricultural base, they direly needed tribute from other city-states to sustain themselves unless they enjoyed natural blessings such as the Nile, whose annual overflow in September provided

Egypt with endless quantities of fertile soil from the highlands of Ethiopia.

Modern urban dwellers tend to feel uneasy about so tentative and qualified an image of the city. We like to think of cities as "eternal" and urban cultural life as free of rural social influences. We are willing to grant that cities do change with architectural innovations, technology, and societal advances but that basically they are immune to the ravages of time and the vicissitudes of history. The tombs of buried cities in Anatolia, Mesopotamia, North Africa, and the ruins of Central America and Mexico are grim reminders that cities can die forever without successors to overlay them, indeed that historically they have been terribly vulnerable to social and ecological changes. Nor do we like to think that cities have any significant roots in agrarian societies, indeed that they did not always "dominate" the countryside in the all-encompassing sense that they do today. The enormous power modern cities exercise over rural life—not only economically and technologically but culturally—seems to be a historic destiny in our eyes that is structured into citification or "civilization" from its very origins, a *telos* that predestines urban life to achieve lasting supremacy over all other life ways. The great urban belts that sprawl across the land, ugly as they may be, appear to us as the culmination, perhaps the fulfillment, of a drama in which early temple cities or Greek *poleis* were merely archetypal characters in a larger and more sweeping civic story that reaches its climax in what we call "modernity" and all its citified amenities. That contemporary urbanization does not represent the culmination or fulfillment of citification, that the history of the city cannot be properly understood with the hindsight and Eurocentricity of a New Yorker, Parisian, Berliner, or Londoner reveals a modern conceit that under critical examination tells us how skewed and misleading are popular ideas of how cities actually developed and the special identities they acquired.

This conceit, to be sure, is not entirely modern. Ancient empires such as the Babylonian and Persian built cities to last and earnestly believed in their immortality. The Greek image of the *polis* as immortal was woven into the heroic image of the democratic citizen and the ideal that noble actions would yield a lasting fame that would survive his own mortality. The words "eternal Rome" are prover-bial. Even before the rise of Christianity, they entered deeply into the consciousness of the city's citizens. As it turned out, the clos-ing years of the empire, a span of two centuries before it finally fell apart, were riddled by deep civic pessimism. The Romans of that time were filled with a fatalistic despair about the future of their city and eventually transferred their allegiance from the "eternal city" to the "City of God."

What distinguished their civic optimism from our civic arro-gance, however, is our preening belief in the autonomy of the city, our conviction that city life and its traditions have always enjoyed a supremacy over rural lifeways and cultures, indeed that the former stands in innovative and exciting contrast to the latter. This conceit is very different from the ancient commitment to the city and has its origins in the highly introverted civic development of Europe during the late Middle Ages. Judging from the comedies of antiquity, Greeks and Romans had their "country bumpkins" who were the naïve foils for "city slickers." But the ancients never doubted that these "bump-kins" were their "country cousins" in a very real cultural and mate-rial sense. The ancestral farm, if an urban citizen owned one, or the family village where the bones of his ancestors lay, was a place to which he repaired from the demands and stresses of city living. It was there that he found his "roots" and from there that he carried his household deities into the very confines of the city. It was from the country, too, that he acquired his bread, cheese, figs, and precious olives so characteristic of Mediterranean fare, for no doubt existed in his mind that, all its cultural amenities aside, the city was mate-rially dependent upon the immediate countryside and the farmers who trundled its produce into his squares.

His political life, such as it was after the decline of the *polis*, had a distinct rural imprint: it centered around assemblies even in Rome and other cities outside Latium, on such councils of "elders" as the Senate, and, in republican times, on a citizen-militia, all of which were institutions formed by early village and town democracies. It is in this sense, by no means a strictly exploitative one, that we can agree with Chester G. Starr's observation that nearly all ancient cities "controlled a wide expanse of countryside, which extended in more recently urbanized areas as far as fifty miles or more from the city walls; they thus resembled an American or English county more than our purely urban units of government. Commerce and industry might have some significance within this complex but did not need to do so, for its unity rested primarily on administrative, agricultural, and psychological bonds."[1] A brisk walk of an hour or so could bring a Roman within clear sight of farmland and within two or three additional hours to villages that seemed largely untouched by urban culture.

That the ancient city also carried enormous political weight in Mediterranean life and was a source of considerable ferment, indeed, potentially, of serious rebellions against its imperial sovereigns, is a fact that never left the minds of the Roman emperors. Not that the peasantry was totally passive and free of strong resentments—not only toward the imperial city but also its own urban centers. Attica's "party of the Plain," as historians call it, enviously regarded the Athenian town-dweller as a parasite and periodically aligned itself against him, hence Kleisthenes's shrewd admixture of urban with rural *trittyes* to neutralize these conflicts. In Roman times a rural police was created by the city, particularly in the Near East, to maintain "order" among the agricultural poor.

But the emperors kept an especially tight rein on the cities under their control, for it was here that internecine conflicts against urban oligarchies and civic chauvinism could reach acute proportions. "The ancient city was an institution which swallowed up much of its citizens' hearts," Starr tells us, and Roman emperors "were undoubtedly right in feeling that the Roman world could not endure outright war among its constituent units but the necessary alternative was deplorable: the race for titles was miserable stuff to fire civic loyalty and to promote urban thinking."[2] Ultimately, through imperial paternalism, legal constraints, and the use of oligarchies to control or deflect the anger of lower orders in their own cities, "the vigor of local politics waned; the cities were not granted, as a substitute, a really significant, inspiriting influence on the imperial government. In all this the Empire was wiping out not merely freedom of thought in the political sphere but the very possibility of serious political action for the bulk of its subjects. So it was destroying a very important focus for the forces which had led to the expansion and remarkable vitality of Mediterranean civilization up to this point. The upper classes and the Caesars both delivered devastating blows at the sense of local attachment, the solidarity of the city, from which the thinkers of the past had drawn their strength." Whatever Roman civilization seemed to produce in the way of order and uniformity of laws, it more than destroyed in the way of creativity and political commitment. By the latter part of the fifth century A.D., the Mediterranean basin had ceased to be either orderly or innovative. In balance, the cost of imperial and bureaucratic control from above with its loss of local control and participation outweighed, by far, the dubious benefits it professed to confer and the entire basin suffered a collapse that was to haunt civilization throughout future centuries.[3]

By contrast to the Mediterranean basin, Europe through the Middle Ages and well into modern times followed a very different development. Although European patterns of civic freedom never equaled the institutional and political creativity of the *polis*, many medieval communes were as vital and energetic as the Greek cities, and it is from this world that modern images of urban supremacy really derive. From the thirteenth century onward, particularly in Italy and the lowlands of modern Belgium and Holland, city-states began to emerge that were structured around uniquely urban tasks—artisan oriented, financial, commercial, and industrial—that slowly loosened urban life from its traditional agrarian matrix and provided the town with an authentic civic life and momentum of its own.

Which is not to say that medieval towns and cities ceased to be dependent upon their rural environs for food and raw materials. But as time passed, in different parts of western Europe, it was to be the city that gradually remade the countryside in its own image and entangled the landed nobility in its own economic and political concerns. And ultimately it was to be the city that imposed its technics, economic relationships, culture, and values on agrarian communities.

Traditional agrarian attitudes did not place a high premium on artisans, merchants, and financiers; these were not orders that were respected in ancient society. A Greek or Roman "bourgeois" who made a profitable income from trade and workshops tried to accumulate enough wealth in order to retire on landed property of his own and live in rural gentility with the airs of a noble. Capital migrated from town to country, not the reverse. The medieval world radically altered this movement. For the first time on a large scale, urban entrepreneurs began to view their vocations as honorable and invest their earnings from generation to generation in their own enterprises.

As family fortunes increased, they remained in the city and were not invested into estates. Quite to the contrary, it was the Italian and Flemish noble who often moved to town and began to enhance his own fortunes by involving himself in industry, trade, and foreign promise. The "merchant prince" of the Middle Ages was not simply a euphemism for men of ignoble lineages who managed to acquire "princely" fortunes from trade. The words often included noblemen of highly prestigious families who regarded the city as a center of a very lucrative commerce and began to participate in it with the avidity of the most pedestrian burgher.

How this reversal of traditional trends occurred can only be noted summarily here. The breakup of the top-heavy, highly parasitic, and costly Roman Empire led to a general retreat of the Latin cities and their European affines in Iberia, Gaul, and Britain to their authentic ecological *chora*: the land base that could properly support a modest urban population. As sources of tribute began to break away from the empire and its provinces were assailed by repeated invasions of semitribal German, Slav, and Magyar peoples, the fragile ancient city dwindled to little more than a village scale, at least in western Europe where urban life had a very limited land base and abutted the vast forest that still covered most of the continent. Rome's population alone declined from its highpoint of a million to less than 50,000 in the "Dark Ages" that followed and never fully recovered demographically through the medieval period, despite its enormous importance as the center of western Christianity. Already, as early as the third century, the European provinces such as Iberia, Gaul, and Britain began to detach themselves from the empire in everything but name and move toward a localist economy that produced only for their own particular needs. As John Mundy and Peter Reisenberg point out, the foundations for a decentralized medieval society were not created by German invaders, the "barbarians at the gates" who sounded the death knell of a terminally ill empire; they were established by Rome itself when the late Caesars, faced with a completely unwieldly imperial structure, set each order in Roman society adrift with the necessary "duties, privileges and special jurisdictions [it needed] to implement its work." Each group, in effect, was obliged to make "its own law" in order to survive. "Moreover, the old classical conception of liberty was replaced by something less grand but more socially exact. Guildsmen, or *collegiati,* were privileged, for example, even to the equestrian grade. But privilege depended upon obligatory service. This perception of an inmixture of liberty and servitude pervaded late Roman and much of medieval social thought."[4]

The veering of the former Roman provinces toward a highly localist society, one in which each component was internally held together by an elaborate hierarchy of clearly defined rights and duties, laid the foundations for European feudalism. Autarchy, now based on hierarchy, surfaced with a vengeance in the form of minutely segmented and self-sufficient communities, all supported by agriculture and a rude form of manorial artisanship. Cities that were once the islands in a rural world largely disappeared while the rural society seemed to solidify into minute congeries of simple manors, each surrounded by small dependent villages. The serfs who peopled these villages were not attached permanently to the land. Both manor and village, in turn, seemed to be haunted by the nearby debris of ruined cities, some partly occupied, which were testimony to an exotic world clouded by legends of past greatness and the misty traditions of imperial power. Within this setting, we do not have to go too far to understand the origins of the Arthurian legends of Britain and the dreams of grandeur that clustered around the name of Charlemagne.

These ruins came to life very slowly, often as the building materials of churches, modest villas, and town halls. The medieval towns and the few cities that we find in the eleventh century after the great tides of "barbarian" invaders and the fractured empires of Charlemagne receded into history were very unusual phenomena. Although they were never independent materially, there is a very real sense in which they were to become quite autonomous economically, to an extent, in fact, that was quite rare in the ancient city. To begin with, they were not simply religious or administrative centers as were so many of their classical predecessors, although such centers were fairly common in medieval Europe. What is significant about them is that they were largely structured around the tasks that most markedly distinguish the city from the agrarian village: crafts, trade, a nascent form of mechanized industry, and a culture—moral, artistic, architectural, and even religious—that is peculiar to such a citified constellation of activities. No longer was nonagricultural work demeaning, nor for that matter was work generally, as ultimately became the case during the imperial period of Rome and the old Mediterranean slavocracy. Quite to the contrary, nonagricultural work was to be celebrated over all other forms as authentic evidence of "civilization," so that, etymologically speaking, the word now came into its own. The artisan and tradesman, normally *metics* whom the ancient city dweller viewed with condescension as the urban counterparts of the rural "primitives" (the view that the imperial age created toward its once idealized yeomanry), now found themselves the center of a new concept of urbanity, indeed, the city's most prized and productive member. Trade itself, which aristocrats at all times slyly exploited to their own advantage, was openly acknowledged as an honorable enterprise, and wealthy merchants began to give lavishly of their fortunes to adorn this new civic temple to their interests. However much the Church viewed profit making as intrinsically evil, it soon learned to accommodate itself to this new civic reality despite its pious denunciations of gain from the pulpit.

This new kind of city appeared not only because of the extraordinary energy Europe brought to a world drained by imperial tribute; it also emerged because of the inherent weakness that marked a highly decentralized and parochial system of agriculture. Whatever may be the image Roman nobles developed of their servile farming orders, the aristocrats of the ancient world still prized their status and traditions as landed gentry. To own a large estate, worked by gangs of slaves and tenants, was as much a calling as it was a lucrative enterprise. The nobles, moreover, had used the Empire to wrest all political power from the artisan and merchant orders, and thus gave their imprimatur to the ancient city: its odium of parasitism, display, indolence, luxury, and class arrogance. The cities they now controlled became monuments to their power as lords of a basically agrarian economy and culture. Under the Caesars, this economy and culture did not disappear; quite to the contrary, it was greatly strengthened by a huge bureaucracy and enriched by an immense flow of tribute. The ancient city never transcended these economic and cultural limits. The *polis*, to be sure, gave it a creativity that made civic life seem like a work of art. Rome, particularly under the Empire, gave it a highly elaborate state apparatus that made it a self-destructive engine for exploiting Mediterranean society.

The medieval city was more fortunate and found itself surprisingly free of these limits. Agrarian society in the Middle Ages was much too fragmented to establish lasting obstacles to urban development:, and its culture was much too introverted to challenge the new "civilization" nurtured by the towns. The famous medieval maxim, "Urban air makes for freedom," could be said to include freedom of thought and an openness to innovation as well as personal and political liberty. Like a calculating observer who can patiently stand by while his opponents exhaust themselves in mutual combat, the medieval towns played for time and waited for the temporal lords to drain their strength in the ceaseless rivalries that marked the era. When the local nobles were weak enough or drained of their wealth, the cities stepped in to assert their own autonomy, either by buying their freedom or by force of arms, and ultimately were to assert their supremacy over the countryside.

This drama, to be sure, did not unfold with the elegant consistency we often expect from a good story. Many cities and towns failed to gain their freedom for generations, despite bitter struggles with their territorial lords. And when they did so, their liberties were often fragile and highly vulnerable to internal as well as external challenges. In retrospect, however, we can see a fairly distinct pattern of civic freedom emerge, not only from territorial lords but also structurally, within the medieval *civitas* itself. The Italian city-states, in particular, provide the *dramatis personae* and almost exemplary "plot lines" for an overall story of European civic development. Not that other cities—Flemish, German, French, and Swiss—are not good examples in their own right, or, in some cases, more lasting patterns of civic freedom. But what Lauro Martines calls the "power and imagination" of the Italian city-states is so vibrant, so artistically creative and institutionally stimulating, that one senses a degree of organicity and roundedness rarely matched for that period by other European regions.[5] And if we rise above the national chauvinism that afflicted medievalists of the recent past, it would be hard to disagree with Mundy that the "distinction between the northern European and Italian town" drawn by Belgium's great historian, Henri Pirenne, a distinction that so greatly biased a whole generation of medieval scholars toward the northern city-states, "seems to have no basis in fact, at least through the ages of urban renewal up to the twelfth century," a time when urban freedom reached its peak internally.[6]

We also know that the movement for civic freedom really began in Italy and only later did it move up to Flanders, northern France, and southern Germany. Thereafter it became a contagion in the cities of western and central Europe. To discuss so mixed and uneven a development in civic freedom, we would be obliged to leap almost dervishly from one region to another to uncover the underlying sequence of events that makes medieval urban development reasonably intelligible. If most of this account focuses on the Italian experience, it is because the peninsula's city-states encapsulate in a clearly definable area the basic cycle that marked the rise and decline of civic democracy in Europe as a whole. Indeed, it was in Italy that the city-state of the High Middle Ages was to come into its own, and it is from the rich accounts of that civic world—its extraordinary art and architecture, its changing political relationships, and its cycle of birth, maturation, and decay—that the European municipality was to play out its most authentic internal development and lead, almost unerringly, to the urban drama that is unfolding before our eyes today.

The origins of these city-states need not concern us. After at least a millennium of urban development, from the legendary founding of Rome in 753 B.C. to the last of the emperors in 476 A.D., probably no area in ancient Europe was more citified than Italy and few municipalities exhibited more staying power than those that the German invaders engulfed as they rolled over the peninsula. Thus, Italian cities never really disappeared. However much they declined demographically, they recovered fairly early and by the tenth century began to grow with exceptional verve and energy.

Nor need we trouble ourselves too much about the factors that led to this recovery. So far as the northern Mediterranean was concerned, the Saracen floodtide that swept over the basin began to recede by the eleventh century: even Sicily was reclaimed for Christianity by 1091, after two centuries of Muslim occupation. By this time, a lively trade existed between Italian and Arab merchants, until, by the twelfth century, the Mediterranean was largely controlled by the Italian city-states. These city-states were soon the major entrepôt agents that brought Christian and Muslim into commercial contact with each other, and the economic foundations for a flourishing municipal life were securely established.

The locale of the Italian city-state, or *communa* to use its medieval Latin name, covered most of the northern and central areas of the peninsula and embraced the great Renaissance cities of Florence, Venice, Verona, Pisa, Sienna, and Cremona, to cite only a few, and those great political nodal points of Italy's history, Milan, Genoa, and Bologna. It is not the Renaissance that will concern us here, except insofar as the Dominican Fra Savonarola represented a popular reaction to the aristocratic and simoniacal world that fostered it, and it is not the *communa* as a "city republic," words so easily interchangeable with "city-state" that will be our main focus of interest. What is insufficiently known about the Italian commune is the extent to which it became a stage for a working democracy and its actors a new expression for an active citizenry. Dazzling as its cultural treasures and adventurous as its stories may be, we must put aside the pure glamour of the commune and examine those brief periods when it seemed more like a *polis* than a republic.

Democracy clearly emerged in the early Italian cities, not only representative forms of governance and oligarchies of various kinds, only to submerge and then reappear again for a short time in richly articulated forms. Its principal institution took the form of a popular assembly and we hear of it almost from the inception of the commune itself. The Italian commune, it should be clear, was not simply a community or even a town in the customary civic sense of the term, although it was this as well. It was above all an association of burghers who were solemnly united by an oath or *conjuratio*. The *conjuratio*, in contrast to the formation of a community by mere force of circumstances and events, turned what we might simply designate as a town into a vibrant fraternity. Although motivated by shared practical concerns, the burghers who took this oath were presumably foreswearing their personal self-interest on behalf of a common one, a conscious act of mutual fealty not to a local noble, cleric, military chieftain, or distant monarch. Despite its strongly religious terms, it was an act of citizenship, unique in an age of religious congregations, that pledged members of the commune to respect each of the others' civic rights and extend them untampered to newcomers and future generations.

The *conjuratio*, in fact, did considerably more: it committed the citizens of the commune to orderly and broadly consensual ways of governing themselves with a decent respect for individual liberty and a pledge to their mutual defense. The stormy conflicts between the monarchs of the Holy Roman Empire and their Italian vassals, and more disruptively between local bishops, nobles, and military captains over landed domains and control over the cities, were resolved by a delineable and, at times, a logical course of urban development to which the *conjuratio* forms a point of ethical entry. Once the cities replaced arbitrary rule with some kind of elective rule, coordinates emerged for gaging its political evolution. Men then swore to aid and defend each other, to pledge their prestige and bring their personal jurisdictions to the service of a shared community, and to respectfully invest the chosen heads of this civic association with executive and judicial powers. A consulate, composed of all the consuls in formal assembly, constituted the highest executive and judicial magistracy of this remarkable association. We know that its members were chosen at a general assembly of the commune itself, a popular assembly that "was quite likely convened with some regularity, and in times of trouble even more often," Lauro Martines tells us. "Here the views of leading men were heard and important decisions taken, usually by acclamation." We know, too, that this general assembly of "all the members of the commune" was the "oldest communal institution" of these Italian cities, and further, that the consuls usually "sounded out the general assembly" before they made any major decisions about such issues as war and peace, taxes, and laws.[7] What is so vexing is that we know so little about this revival of a highly democratic, face-to-face form of political control and of the citizens who created it. Whether age endowed it with the legitimacy of an unbroken tradition back to ancient times, we do not know— only that it was the oldest of the commune's structures. This much is clear: whatever authority the assembly enjoyed prior to the consulate—that is, if it existed at all before the *conjuratio*—is clouded by the furious infighting between clerics, nobles, opportunists of all kinds who managed to grasp and lose control of the cities for nearly two centuries before these civic entities freed themselves of arbitrary rule and disruptive rivalries.

Considering the volatility of the Italian city-states, the assembly soon lost ground to the consulate and the consuls, who began to elect their own successors. The Italian city was never to be restrained by such Hellenic mores as "nothing in excess" or the Pythagorean ideal of *harmonia*. One could almost say that seldom did anything but excess and explosive political theatrics mark these fervent and creative urban entities. In any case, civic leadership fell into the hands of large, wealthy families who were to establish the parameters of civic politics for centuries to come, even after they were temporarily removed from power or replaced by newcomers from within or outside the city. The assembly, in turn, never totally disappeared. It lingered on in a highly atrophied form and haunted the oligarchs and despots who usurped its powers. We know that in Bergamo, for example, as elsewhere, outgoing consuls were obliged to hold elections for new ones in popular assemblies, although the nomination of candidates was the privilege of only a few. Such practices persisted well into the twelfth century when cities such as Genoa and Pistoia tried to eliminate prearranged nominations of consuls between electors that favored one of the parties at the expense of another.

In time, this early popular democracy gave way to republican forms of governance, and the powers of the city were invested in a large legislature dominated by influential urban families. In some places, the assembly continued on as a vestige to be manipulated by oligarchs. Elsewhere, it was replaced by "assemblies" that were simply republican legislatures. "Emulating all princely courts and the general assemblies in Italian towns, some of these [representative] assemblies hoped to meet four times a year," observes John H. Mundy. "This was impracticable for general assemblies, but smaller select councils often did meet three or four times yearly"[8] Although Mundy is referring here specifically to England, the drift from popular assemblies to oligarchical councils was as true of Italy as it was for Europe generally. The "conciliarism," as this parliamentary trend was called, essentially replaced civic democracies, such as they were, with authentic city republics, and the great towns of Italy looked more, institutionally speaking, like preimperial Rome than classical Athens.

We must not suppose, however, that Italy became a centralized state at this time, even on a municipal base. Whether as a democracy or a republic, the Italian city rested to an astonishing extent on little neighborhood "communes" that existed within the larger urban commune. We have no parallel for this development in ancient Athens or Rome. Irrespective of the kind of consulate it had, the city's neighborhoods acquired an autonomy of their own that is truly spectacular by any standards of civic governance. Fortified towers rose everywhere—in Milan, Florence, Pisa, Verona, Pavia, Parma, and the like—held by noble families in *consortia* or sworn family groupings that were pledged to maintain their neighborhood turf against other corporate intruders and the commune as a whole. Oaths were sworn everywhere in the tradition of the *conjuratio* but so localist in their commitments that we can only describe a commune as a loose confederation of neighborhood communes, indeed, a confederation of neighborhoods within the commune. "When a union of this sort brought peace to a neighborhood," observes Martines, "that part of the city became nearly impregnable, because the linking of towers and the easy blocking of the narrow streets made for an effective domination of the adjoining area."[9]

Let there be no mistake about the fact that these *consortia* were anything but neighborhood democracies. The towers were built by leading wealthy families who were much too powerful in some respects, and much too weak in others, to seize complete control of the commune. They were highly localist oligarchies that kept Italian cities in continual turmoil until they were supplanted by despotic *signori*, family dynasties ruled by one man who replaced a multitude of oligarchies, often to be unseated again and his sons restored in reckless shifts of aristocratic and military power. Italian civic history is not notable for any even or linear development from democracy through republican systems to oligarchy, monarchy, and despotism. All five of these institutional forms bubbled up in an unpredictable assortment of patterns, predominantly as one at certain times or in combination with each other on many other occasions. Ultimately, the Renaissance cities settled down as oligarchies and princely courts where an electrifying but aristocratic humanism gave us the priceless artistic treasures of that era.

But the *consortia* created "bad" and lasting habits that fostered a highly democratic atmosphere at the base of city life. It produced neighborhood loyalty and citizen empowerment with a vengeance. "Propertied urban inhabitants were attached tenaciously not merely to a city but to a street, a parish, an ambience—to a radius of perhaps 150 meters," declares Martines. "And though a citizen might frequent an estate in the country, he lived in the city and there, if fortunate, he died. More likely than not, he married someone from that vicinage, and all his main blood ties, as well as his closest friendships, were there. The character of the neighborhood could be so distinctive, could leave so deep an imprint on its inhabitants, as to make for different linguistic expressions and even for different intonations or accents. Dante detected neighborhood differences in speech at Bologna."[10]

These loyalties gave the commune a political vitality it never fully lost until recent times. Conciliar governance did not abolish major democratic features of the city such as the selection of consuls by sortition at Lucca, for example, where the neighborhoods or *contrada* convened in their own assemblies and "lots were drawn," notes Daniel Waley; whereupon "each of the 550 men who drew slips inscribed *elector consiliari* then had the duty of naming one man from his own *contrada* as a councillor."[11] When the *podesta* system of government or *podesteria* was established in Italian communes that vested all executive power in a single man, he too was often chosen this way. In Vicenza during the mid-thirteenth century, twenty city electors "were chosen by lot, and of these twelve were eliminated by voting; the eight remaining electors then proposed three names, from which the final choice was made by a further vote of the council."[12] The councils, in turn, often required a quorum of two-thirds to arrive at a decision, and in some cities councillors were fined for failing to attend. Nor is it unusual to learn that a majority of two-thirds was needed to make major decisions, at which time even larger quorums and majorities as high as 10/11 or 16/17 were required—proportions that could almost be regarded as consensual.

The *podesteria*, for all its majesty and seeming absolutism, was by no means the authoritarian ogre it initially appears to be. A *podesta* was not selected from any members of the community. He had to be an outsider who could distance himself from the internecine familial and factional conflicts that often brought the city to the brink of chaos, hence too the elaborate electoral system that was meant to buffer the influence of particularly powerful groups. Nor could he come from a neighboring commune, which obviously was privy to the city's infighting. He was selected for his impartiality, for his legal training, and ordinarily held his office for no more than a year, often as little as six months. The consulate, like the assembly, did not disappear; indeed, in Treviso as late as 1283, general assemblies of virtually all resident citizens who enjoyed rights to public office were convened to deal with emergencies, even if only to bestow almost despotic powers of control on a military captain.

A detailed account of the *podesteria, signori*, the later courts of the Sforza and Medici, and the republican oligarchies that overturned these various forms of one-man control, would require a volume in itself. Here we must focus on the conflict between the "people" and the "nobility," the *popolo* and *nobilta* as they were called respectively. Equivalently, we could have called the "people" the "foot soldiers" or *pedites* and the "nobility" the "knights" or *milites*, the military characterizations so redolent of the rise of Greece's *hoplites* and the supremacy they achieved over the aristocratic cavalry. The parallel, limited as it may be, is quite instructive. What the Italian citizens of the twelfth and early thirteenth centuries called the *popolo* was in no sense the urban poor, the *thetes* of ancient Athens. The poor in the commune were a socially inchoate stratum of journeymen, relatively unskilled craftsmen who verged on being proletarians, laborers, servants, itinerant tradesmen, runaway serfs, and that vast *déclassé* of beggars and thieves who plagued the medieval cities almost from their inception. Many of these elements could be expected to follow any standard that filled their empty pockets. In a feudal world where a feeling of social place was crucial to one's sense of self-regard, they were outside the pale of the commune's life and its broader political concerns.

What the commune meant by the *popolo* were men with a certain amount of material substance, such as master craftsmen, professionals, notaries, well-off tradesmen, even financiers and an emerging commercial bourgeoisie that owned sizable fortunes from foreign trade. Although there was a close interlinking of the *nobilta* and the very well-to-do *popolo* that makes it difficult to draw precise class lines in the conflicts that brought the two into opposition with each other, many of the *popolo* were virtually excluded from the city's political life and were treated like resident aliens, the *metics* of ancient Athens. They paid its taxes, served in its militia, and were subject to all its laws without any right to hold public office or enter into its civic councils. In cities of 40,000 or more, active citizenship was restricted by property qualifications, social status, and length of residence to a thousand or less. Although the Milanese *popolo* provided the city with the greatest part of its revenues, for example, they held only one-fifth of the consulate's offices. Later, on the intervention of the Holy Roman Emperor, Otto IV, they were given half of the conciliar seats. This reform and others like it were enough to whet appetites without providing enough sustenance to keep them satisfied. In the first half of the thirteenth century, the *popolo* began to take over the reins of power in one commune after another—Bologna by 1231, Pistoia in 1237, Florence in 1250—and to acquire a growing voice in the governance of Piacenza, Lodi, Bergamo, Siena, Parma, and Genoa. "The *popolo*'s breakthrough into politics was the result of revolutionary organization," declares Martines with a decisiveness that is not easy to encounter among many historians of the era.[13] In a very real sense, perhaps more sweeping than we realize, he is correct.

In retrospect, what makes the *popolo*'s ways of organizing so revolutionary was its highly localist and organic fashion of acting politically, a form of organizing that is utterly antithetical to modern concepts of party politics. The *popolo* had turned themselves into a neighborhood movement, not unlike the *contrada* in structure and the *consorterie* in their confederal way of interlinking with each other. The noble families of two centuries earlier had taught its more servile orders well in the arts of political organization and power. The tradition of organizing a movement on a block-to-block, square-to-square basis had not died out in Italian cities after the *podestarie* subdued the interfamilial squabbles within the nobility. It was still very much alive, and the *popolo* used it well. However varied the specific structures this organization produced, certain forms clearly emerged from this neighborhood base.

The rise of vocational guilds provided one very important pattern for linking men of the same occupation from different neighborhoods into a common organization. Guilds were established not only for merchants, physicians, apothecaries, jurists, and notaries, but also for smiths, cloth finishers, butchers, bakers, furriers, tanners, leather makers and the like, generally reflecting the *popolo*'s social composition as a whole. These vocational groups, composed almost exclusively of well-to-do tradesmen and master craftsmen in their fields, were the earliest popular organizations of which we have any record, and they tried not only to control the operations of their crafts and its produce but also constituted themselves into armed societies with the very distinct goal of controlling the commune. Before long, open clashes exploded between the *popolo* and the *nobilta* at Brescia in 1192, Piacenza in 1198, Milan in 1201, and in year-by-year sequence, Cremona, Assisi, Lucca, and onward to Pistoia in 1234.

Neighborhoods were now convulsed by strife between poorly trained and the more highly skilled fighters of the noble *consorterie*. Rarely had the armed citizen moved so much to the forefront of civic political life. By now, armed popular societies based on neighborhood ties began to supplant the guilds in order to deal with the endemic warfare that erupted in the city's streets. These societies were better trained than the vocational organizations. More closely cemented by their ties as neighbors, they had a better knowledge of the quarters in which they functioned and they could muster their forces more rapidly and on shorter notice as emergencies arose. If Bologna can be taken as an example of the enhanced ways of dealing with the *nobilta*, it demonstrated the need for the armed neighborhood societies to extend beyond the guild framework into a more cohesive, area-oriented, well-synchronized, and highly cooperative form of military organization. By the middle of the twelfth century, the city could boast of twenty-four armed neighborhood companies in comparison to Florence's twenty and still smaller numbers elsewhere. Moreover, the Bolognese societies were consolidated across adjacent urban parishes into what looked very much like a citywide militia. The *popolo*, to be sure, had not created the idea of a neighborhood militia out of thin air; such companies of men living in the same parts of the city had been the backbone of the communal military system for years. What the *popolo* did, however, was to change the militia's function from a mere civic guard, designed exclusively to defend the commune from external enemies, into a politically insurrectionary force designed to change the commune's internal structure and power relationships.

Florence's armed societies can be taken as a general example of the structure that the *popolo* established in the popular communes of Italy after their victory over the *nobilta*. The twenty neighborhood companies it had created earlier now became the commune's exclusive fighting force, absorbing the remains of the old one where it could. In districts where the *popolo* was weak, a larger number of militia units were assigned to guard its interests. The rest were equally distributed throughout the city. Each company adopted its own banner and insignia—a "dragon, a whip, a serpent, a bull, a bounding horse, a lion," even a ladder, "were emblazoned on individual shields and helmets," Martines notes. "Rigorous requirements required guildsmen to keep their arms near at hand, above all in troubled times. The call to arms for the twenty companies was the ringing of a special bell, posted near the main public square. A standard-bearer, flanked by four lieutenants, was in command of each company. These men were elected by a council of twenty-four members of the company and served one year. The Captain of the People was the commander of the twenty companies. Being an outsider, he was expected to stand above the temptations of local partisanship. His oath of office obliged him to call the companies to action whenever there was a threat to the *popolo*.... The *popolo* had no trouble finding only the most committed and reliable men for its armed companies. The success of this was seen in moments of crisis, when the ardor of the companies won many victories for the *popolo*," a fact that was not lost on Machiavelli some two centuries later when he voiced his compelling argument for a citizen-militia as opposed to the use of mercenaries in *The Prince*.[14]

Once it came to power, the *popolo* did not completely eliminate the podestal constitution. A large Council of the People was formed to counterbalance the power of the old communal legislature and a powerful council of "elders" functioned as the commune's executive. The Captain of the People was elevated to a status that, for all practical purposes, matched that of the *podesta*, and he was fully regaled with a court system that paralleled that of the podestal one. The *popolo*, in effect, had structured a dual power into the older constitution, matching the communal council with a popular one, the *podesta* with the militia chieftain, and courts of both men as counterweights to each other. Overarching this entire structure was the popular militia itself, which could easily intervene in the *popolo*'s interests if its interests were in any way threatened.

If this was a democracy, it was one of merchants, notaries, professionals, and master craftsmen, not of journeymen, weavers, dyers, and laborers who formed the only "proletariat" the thirteenth-century world knew. By the same token, however, it was not a democracy of blooded idlers and aristocratic bullies. The popular democracy was composed of men who worked with their hands at crafts or devoted long hours of the day to the chores of administration, business, and legal affairs. Whether they were wealthy entrepreneurs or men of modest means, they saw themselves as self-made individuals, not the heirs of family wealth and unearned status; they believed, with good reason, that they lived highly productive, sober, and socially useful lives and were far more capable than the nobles of bearing serious responsibilities on their shoulders. They were no one's clients and no one's masters, except where they trained their own successors in the skills of their crafts. Although the new manufacturers who operated with hired labor could hardly make this last claim for themselves, they were far from constituting a majority of the typical Italian *popolo*. Italy in the thirteenth century was not an industrial society with modern capitalists and workers. It was a largely artisan and highly paternalistic world, and the men who fathered its workshops as well as its families were not disposed to grant the least modicum of power to anyone who was a subordinate. These men knew only too well that, without them, the commune would quickly atrophy and disappear, but without the nobles and, to a considerable extent, their own subordinates, it could survive and ultimately flourish.

For limits of space, it is impossible to explore the vast cultural impact the popular commune had on its age, be it the painting of Giotto, the sensual realism of Boccaccio, the shift of literature from Latin to the vernacular, the laicization of education, the huge increase in popular literacy, or a political tradition that trounced any attempt at institutionalized despotism. For all his aristocratic inclinations, Dante was a *popolani*, no less than the chronicler Giovanni Villani whose history of Florence is one of the best to be written of his era. The popular commune created new men who breathed a fresher urban air than any of their medieval contemporaries: one relatively free of superstition, turgid scholasticism, and social status. Above all, it produced a new kind of citizen who, with his weapons by his side and craft tools in his hands, felt boldly empowered and politically confident. Alert to any threat to his liberties, he was prepared to fight for his rights with the same energy with which he fashioned his produce. This alertness extended beyond a defensive stance against an aggressive nobility into a militant interest in public affairs and a readiness to occupy public responsibility. Not since Hellenic times did men of relatively modest means feel so close an identification between their own interests and those of the city in which they lived. Deeply individuated, indeed animated by a vivacious love of public debate and political discussion, such as Italians are to this very day, this new citizen seemed all the more motivated to shape his city's destiny as well as his own. The ancient *civitas* as an association of active citizens had been reborn, and its spirit was to live in its citizen body long after its institutions were to atrophy and pass into history.

That the popular commune did not last as long as the Athenian *polis* is no mere quirk of ill fortune. Like the Roman Republic, its problems could not be resolved by replacing a small oligarchy with a bigger one. The popular commune, was not a "leveler" of social orders and economic classes. Quite to the contrary, it gave a new impetus to trade, the accumulation of wealth, and increasing economic and political differentiation, setting the *popolo grosso* or "fat people" against the *popolo magro* or "thin people." In time, the commune developed a siege mentality, fearful of the *popolo magro* within its walls and the *nobiltas* who were slowly gathering their forces outside them. To live with one's weapons by one's side with an ear open to the requests of a buyer and another to the peals of an alarm bell is not conducive to trade and is exhausting to the spirit of a nascent bourgeois. After a few generations—and the time span varied from city to city—signorial rule began to overtake republican virtue. The process was as insidious then as it is today: popular assemblies began to vote increased emergency powers to their *podesta* and captain; the same men were repeatedly confirmed in executive office, and their tenure was extended from one year to five, from five to ten, and ultimately to lifetime office. Soon the leader's son supplanted his father after death and dynasties began to turn elective offices into mere charades. However much they lingered on ceremoniously in form, the republics began to die in fact, and the Italian cities, unevenly and with trepidation, entered the age of the *signori*. That the republican era never ceased to haunt the Italian mind is attested to by sporadic attempts to revive it. The most notable, to be sure, were Cola di Rienzi's assumption of power as the tribune of a new Roman republic in the mid-fourteenth century and Fra Savonarola's theocratic republic in Florence toward the end of the fifteenth. Rienzi, however, had his eyes focused more on a unified Italy than an emancipated Rome, a project that was woefully advanced for its time, and Savonarola turned Florence into a hysterical battleground for an apocalyptic battle between Christ and Satan. If Rienzi was too early for the fourteenth century, Savonarola was too late for the fifteenth. Both men challenged entrenched attitudes that stood as insuperable obstacles to their projects: Rienzi, a predominantly feudal sensibility; Savonarola, a vivacious Renaissance sensibility that increasingly made his monkish mentality an anachronism. And both men went down in tragic defeat: Rienzi as a hero of Italian nationalism, Savonarola as a fanatic of Italian medievalism.

There can be no doubt that all of Europe watched the career of the Italian city-state with keen interest and borrowed heavily from its pioneering civicism. If Europe had any capital during the High Middle Ages, it was papal Rome, also an Italian city-state. Pilgrims and invaders, kings and beggars, knights and merchant-adventurers, all crisscrossed the peninsula for reasons of their own, and Italian merchants were to be seen everywhere: at fairs in France, city gates in Germany, on London's docks, and in Flemish city squares. Hence, if civic freedom needed a stimulus it could not generate on its own, the Italian city-state was a ubiquitous presence that could provide one. In any case, and in almost every way, the cities of western Europe went through all the phases that the Italian cities had pioneered and added certain adornments of their own.

Generally, European towns north of the Alps lagged behind the Italians by a half-century or so, and at some point in time they branched off from this shared development into institutional stagnation. Various kinds of republican structures can be found at one time or another in almost every town of Europe, often preceded by consulates and followed by oligarchies with a wide variety of names. Elective systems of governance, which Roman traditions inspired in Italy, were reinforced north of the Alps by German tribal traditions. Not until well into the High Middle Ages did hereditary kingships come into fashion; until then, monarchs were elected to the throne and by no means did they occupy it with complete regal confidence. Moreover, they were surrounded by councils that spawned the parliamentary structures of later times, essentially oligarchic in character but in some measure reflective of the aristocratic, clerical, and commoner estates that made up the sovereign's realm. As in Italy, the word "people" was defined in a highly restricted manner. "Note that there is a difference between the people (*populus*) and *plebs* which is the same as that between an animal and a man or between a genus and a species," declared a thirteenth-century canonist in the Aristotelian language of the day. "For nobles and non-nobles collected together constitute the *populus*. The *plebs* is where there are no senators or men of consular dignity."[15] The inferences we can draw from these remarks are that the "people" include everyone of all orders and degree; the *plebs* consist of that segment of the "people" who are not equipped to govern. The governing councils of European towns generally reflected this distinction: an urban patriciate monopolized most of the civic offices. Rich merchants took over Bruges in the mid-thirteenth century; Rouen in the late twelfth; Freiburg, which established a lifetime consulate, in 1218; and many large cities of Germany such as Cologne. The conciliar system and the early assemblies were reworked into oligarchies. Needless to say, there were also exceptions. At Nimes, the constitution provided for the election of five consuls from the "whole people of the city or its larger part," typically indirectly, such as existed in the Italian city-states. Arles locked

its carefully controlled general assemblies into an oligarchy by placing virtually all changes of law and imposition of taxes in the hands
of its consuls and their council.

"Democratization" of the European cities occurred slowly and
generally followed on the heels of the popular communes in Italy,
often in exacting detail. Basel established a captain of the people in
1280; Freiburg turned its oligarchy into an annually elected board
of twenty-four consuls following a "popular" revolt very similar to
Bologna's; Liege created what can best be called a guild-type city
republic, headed by sixty governors of its craft guilds and two bur-
gomasters. The city went even further than most when, from 1313
onward, it made the issuance of new laws contingent upon the approval of a popular assembly composed of all the city's citizens,
irrespective of their status. Even broader extensions of civic liberties were to mark the stormy popular movement of Flanders when
civic government was shaped by the despised weavers and fullers
of cloth-manufacturing towns such as Ghent and Ypres where the
"fat" and "thin" people were to clash in brutal conflicts. This chronic
struggle is too complex to narrate here; it involved not only battles
between well-to-do oligarchies and virtual proletarians who manufactured wool products as mere hired help but also the intervention
of the Count of Flanders, Guy de Dampierre, and the French monarch, Philip the Fair. Patriotism intermingled with class conflicts to
a point where democratic demands were repeatedly diluted by the
hatred of the Flemish people toward France. As a result, lowly artisans and their lordly masters fought side-by-side when they were
not fighting face-to-face. Ultimately, the "thin" people, organized
into "lesser guilds" triumphed over their patriciates, and established
a civic structure that gave adequate representation to the weavers,
fullers, and guildsmen of "low degree" in a tripartate magistracy that
excluded all or most of the patricians.

What the European communes could not resist was the steady
rise of the nation-state, indeed one that the rich merchants favored
to restrain their unruly local barons who interfered with the free
movement of trade and their unruly laboring classes who posed a
continual threat to civic oligarchies. Perhaps the earliest harbinger

of this new drift toward territorial centralism occurred when the Flemish communes, led by Philip van Artavelde, the son of the great champion of Flemish liberty, Jacques van Artavelde, were thoroughly routed by Count Louis in the Battle of Roosebeke in 1382. "Henceforth, Flanders was to give up the dream of government by a league of independent towns and submit to the ever more and more centralized rule of powerful territorial lords, whose model was naturally the aggressive monarchy of France," concludes Ephraim Emerton in his wistful account of this tragic history. "At the death of Count Louis in 1383 the county passed by inheritance to his son-in-law, Philip the Bold of Burgundy, the first in the line of princes through whom the Low Countries were drawn into the politics of France, the Empire, and Spain. The cities, still great and powerful, enjoying very wide privileges, were to be only incidents in the larger relations of the country."[16]

Emerton's closing remarks on the struggle of Flemish towns for civic freedom and autonomy might well seem like a touching epitaph for the commune as the nation-state began to emerge. Not only did it survive as little more than a haunting dream in Flanders, but also in Italy, where the triumph of the *signori* by the fifteenth century was virtually completed. Savonarola's brief restoration of a Florentine republic in the closing years of that century reads more like a parody of the city's glorious republican past than an authentic revival. It would seem that Daniel Waley's verdict on the nature of Italian civic republicanism could be applied to European communes as a whole. "The existence of republican forms of government in the Italian city-states was intensely precarious," Waley writes in an assessment that can only be regarded as an understatement. "These institutions were so constantly under pressure or even in full crisis that there is little reason to be puzzled at their failure to survive in most of the cities."[17] Internal factionalism, which often reached insurrectionary proportions, "suffices on its own to explain why, in most cities, the regime of a single individual was able to secure acceptance before the end of the fourteenth century. Clearly the occasional survival of republicanism as an exception needs more explaining than do the triumphs of the *signori*. Republicanism in decline is seen as a puzzling historical problem only if the republic is regarded as the body politic in health, the 'tyranny' (*signoria*) as a pathological form."[18]

Perhaps, as Waley seems to suggest, we should reverse the judgement modern liberal historians have of what constitutes the health and pathology of a body politic. But it is fair to ask, as I have throughout this book, whether any kind of politics, much less one that can be "embodied" in an active citizenry, is possible under a tyranny or, for that matter, a "representative democracy" (republic). Waley seems to surrender the need for an ethical judgement of the commune, of the *conjuratio*'s radical implications and its sense of promise, if he asks us to surrender to the "facts" of the historical record rather than cherish the possibilities civic freedom holds for a humanity that has yet to fulfill its unfinished potentialities for a rational way of life. We can never let this commitment to an ethical judgement give way to the evidence of a grossly irrational reality merely because the record of the human enterprise speaks faultingly against it. Truth rarely lies in the center of experienced facts. If it did so, we would still believe the earth is flat and justice is merely the authority of the strong. These are the irrefragable facts as they are experienced in everyday life. Historians, in turn, would merely be turgid chroniclers of anecdotes, storytellers of things past whose accounts can never be free of their own hidden biases and proclivities. Until very recent times, philosophy for more than two millennia has warned us that "brute facts" do not really exist and are never free of interpretation, even as we deny humanity any sense of possibility, the hope of going beyond the "facts" and achieving its most worthy ideals. "Brute facts" will become real only when we become brutish and deal with the existing reality on its own terms, no different in principle from the adaptive ways we impute to other living beings.

Actually, the history of civic freedom does not end with the fourteenth century or the rise of the nation-state. The ideal of popular assemblies as a form of self-governance, and the city as the nuclear arena for politics and an active citizenry, seems to have enjoyed a remarkable degree of persistence, especially if we bear in mind that

there were so many factors—statist, economic, logistical, and demographic—to work against. That it exists even today in the form of the
New England town meeting, however vestigial it appears by comparison with its past, is testimony to the need people have felt in fairly
politicized regions to retain forms of face-to-face decision making
and direct democracy. In fact, the town meeting is much too close to
our discussion of the city's future to be examined in any detail here.
It may be regarded as a legacy, but it is still a living legacy that should
not be consigned to the past. Suffice it to say, for the present, that
from 1760 to the closing years of the American Revolution, it spread
out of New England and reached as far south as Charleston. During
this period, it fueled the revolution as surely as the more well-known
Committees of Correspondence and such popular societies as the
Sons of Liberty. Considerable effort was needed to eliminate the
town-meeting form in many communities of the new United States
and to nibble away at its powers in New England itself. We shall see
that as recently as the 1960s, Vermont townships and their popular meetings essentially controlled the state government because
the House of Representatives was elected by towns, not along geographic and demographic lines.

Nor can we deal with the challenging problems raised by the
Swiss Confederation, a piece of contemporary history that seems to
have been shunted away from "mainstream" accounts of civic freedom. To assign a few paragraphs to a living illustration of confederalism, participatory democracy, and active citizenship, despite the
philistine smugness of Swiss life and the conservatism of Swiss politics, would be doing violence to the extraordinary example of freedom that is also with us. It will require a section of a later chapter
on civic confederal forms to even remotely approximate the insights
Benjamin Barber brings to this issue in his discussion of direct democracy in the Swiss cantons. What Barber asks us to recognize is
that democratic forms in themselves, however radical they may seem
to be, "are inherently neither conservative nor progressive; they can
be employed to bring about rapid and even radical change under
conditions of consensus and unity of governing and people. They can
obstruct reform and paralyze effective government altogether where

the people are no longer one with their governors or synonymous with their government.... Democracy in Switzerland appears reactionary because the conditions that justify it have eroded; democracy in the villages of Graubünden is failing because the village communities are themselves moribund. Neither the institutions themselves nor a putative peasant mentality is to blame."[19] Democracy, in sum, has its own prerequisites, and it will be no small part of the coming chapters to examine them in the light of modern societal problems— American, European, Graubünden, no less than those that exist for a New England town-meeting democracy.

The most dazzling, almost meteoric example of civic liberty and direct democracy in modern times is the rise and brief ascendancy of the sectional assemblies in the Great French Revolution, a movement that addresses the typical question of whether or not a face-to-face democracy is possible in a large city, indeed by the standards of two centuries ago a world city. It should be kept in mind that eighteenth-century Paris, where the sectional assemblies flourished, was the capital of continental Europe's most absolute monarchies, a city that, more so than any other aside from London, seemed to have been a creature of the nation-state. Under the Bourbons, Paris had very little self-government of its own. Ruled more or less directly by the monarchy, it groaned under the imposition of an inept regal administration and a corrupt nobility that reduced its lower orders to sheer destitution. Yet within a span of only four years, Paris acquired not only complete control over its own civic affairs but created a system of control that was built around face-to-face neighborhood assemblies, coordinated by a commune that, at its revolutionary highpoint, called for the complete restructuring of France into a confederation of free communes. This demand was to survive for nearly a century after the revolution, when it became the insurrectionary program of the more famous Paris Commune of 1871.

Ironically, the sectional assemblies have their origin in the sixty district assemblies Louis XVI convened in Paris to elect the city's middle-class deputies to the Estates General. These assemblies were directed to select the 147 electors who, in turn, were to choose Paris's twenty deputies to the Third Estate. The French Revolution, although spurred into action by the power plays between the Third Estate and the monarchy, really began from below and, most typically, as a result of Parisian militancy. Despite the highly discriminatory franchise that the King imposed on his own capital—nearly a quarter of the city's population was disenfranchised in contrast to a sixth in the rest of France—the district assemblies were highly spirited and acted in flat defiance of the throne. They rejected the presiding officers with whom they were saddled by the municipality and chose presiding officers of their own; the 147 electors to whom they were entitled were unabashedly raised to four hundred; and what is perhaps most damning, they refused to go home as the monarchy directed them to do. They defiantly established themselves as permanent fixtures of Paris's municipal government. The four hundred electors, in turn, began to meet almost daily at the museum and shortly afterward at the Town Hall, where they simply edged out the royal appointees to the municipal government. The electors, in effect, became a provisional Paris "Commune"' until an authentic Commune was constitutionally elected. By September 1790, the Paris Commune numbered three hundred and coexisted uneasily with the sixty district assemblies from which it drew its representatives.

The tension between the Commune and its district assemblies and within the assemblies themselves mounted as the revolutionary fever of the city rose. The distinction between propertied "active citizens" who had franchise rights and propertyless "passive" ones who were expected to be little more than political bystanders ultimately made any exclusion of the disenfranchised people, the *sans culottes*, increasingly awkward. By this time, the "districts" had been renamed "sections" and their number reduced from sixty to forty-eight by the National Assembly in order to diminish their power. Despite such attempts to suppress the "districts" and later the "sections" as deliberative bodies, the assemblies largely ignored the National Assembly's decrees. In May 1790, the Assembly tried to limit the concerns of the sectional assemblies to purely municipal issues, a ploy typical to this day of conservative elements in New England town meetings, but the restriction ultimately proved to be ineffectual. With the outbreak of war, the revolution moved markedly leftward. In July 1792, the *Theatre-Français* section decreed the elimination of all distinctions between "active" and "passive" citizens, a decree that was quickly adopted by all the forty-eight sections of Paris. The sectional assemblies were now thrown open to all the underprivileged orders of Paris: the laboring classes, relatively unskilled artisans, carpenters, construction workers, nascent proletarians of the textile and luxury goods workshops, and the like. A year later, on Danton's request, the newly elected revolutionary Convention that replaced the National Assembly decreed that poor citizens could receive a stipend of forty sous for attending sectional assemblies, which were reduced to two a week, presumably so that the *sans culottes* could attend them with reasonable regularity. The attempt by the liberal Girondins to close all sectional assemblies by nine in the evening so that working *sans culottes* would be unable to attend the assemblies after they returned home contributed greatly to the popular insurrection that removed them from the Convention and brought the Jacobins to power.

Although the details of how these local assemblies functioned leaves much to be desired, partly because the minutes of the sections were lost in the flames of the Paris Commune of 1871, partly, too, because their history has yet to be written by historians sympathetic to their decentralistic outlook, we know enough to describe them in broad terms. Sectional assemblies met at least twice weekly in the summer and autumn of 1793, the high tide of the *sans culotte* movement and often went into "permanent session" during hectic periods of the revolution. Attendance fluctuated widely from a hundred or less when the agenda was routine to overflowing halls (usually in state-commandeered churches and chapels) when serious issues confronted the revolutionary people. Structurally, each section had a president with a committee to assist him, a recording secretary, tellers for counting votes, and ushers to maintain order. Although it is not very clear how fixed these positions were, the President's committee was renewed monthly by an assembly vote, and very few presidents seem to have held office for more than a year. Each section, it must be stressed, made its own rules and probably varied in form and the frequency of its meetings.

A multitude of revolutionary commissars and committees surfaced from these assemblies and dealt with a large number of important economic, political, and military problems. We have only to enumerate the names of the sectional committees to gain some idea of the functions this neighborhood democracy undertook. Aside from its *assemblee generale* of the local citizen body, each section's "most important function, that of police, was entrusted to a *commissaire de police*, aided by sixteen *commissaires de section*. To meet the needs of local administration the sections formed almost as many committees as a modern American town or city. There were *comités civile* and *comités revolutionaires* (Vigilance Committees); there were *comites de bienfaisance* (Relief Committees), *comités militaires*, *comités d'agriculture*, and *commissions de salpêtres* (for providing gunpowder). Each section had its own revolutionary court system and justices of the peace as well as special committees to organize work for the unemployed (*ateliers de charité*), or tenth-day festivities (*fêtes décadaires*), or open-air suppers for the poor (*banquets populaires*).[20]

The forty-eight sectional assemblies, in turn, were coordinated by the Paris Commune to which each section elected three deputies at an *assemblée primaire*. Also at this special assembly, usually quite heavily attended, the sections elected the *Bureau* of the Commune: the city's mayor, its *procureur*, and his two deputies, who essentially constituted a communal executive committee to which sixteen *administrateurs* were chosen from the Commune's normal complement of 144 sectional representatives. To buffer this rather manipulative executive from popular pressure, an additional thirty-two members were added to the *Bureau* to form a *Corps Municipal*. When this majestic group of forty-eight sectional notables convened with the remaining ninety-two deputies that formed the Commune, it bore the solemn title of the *Conseil Général de la Commune* or, as it came to be known in the historical literature, the Commune's General Council.

The more radical sections of Paris were not unmindful that these concentric rings of authority surrounding the commune's Bureau were meant to camouflage an inner circle of executives who were as disquieted by the irascibility of the sections as the Convention was of the Commune, a body that it really saw as a dangerous challenge to the nation-state. Although the Commune eventually became more radical under sectional pressure as time went by, the sections were always more radical than the Commune and began to act very much on their own. When necessary, as was frequently the case, *sans culotte* sections did not hesitate to bypass the Commune completely and form their own interlocking committees and networks. Delegations from one section often visited the assemblies of another, where they were usually greeted with fervent embraces, hortatory speeches of welcome, and characteristic "kisses of fraternity," which often culminated with banquets topped by endless toasts to liberty and the brotherhood of man. As F. Furet *et al.* tell us: "From the angle of political solidarity, the participants in activities of the sections saw themselves as forming a whole, a 'family' of 'brothers and friends,' where the use of the familiar form *tu* broke down all the barriers of social and cultural origin. Outside the section this fraternal solidarity was further expressed by 'fraternization,' a sort of pact for mutual aid between the popular sections to resist the pressure of the aristocrats and the moderates, and then to eliminate them from the committees which they controlled. Thus on May 14, 1793, the Social Contract section, under the direction of its president, went in a body to the Lombard section to chase out the 'aristocracy.' In this way the sections were regenerated in the bourgeois quarters dominated by the moderates before the fall of the Girondins, and by the 'new moderates' after the month of September 1793."[21]

It is difficult to fully convey the enormous powers the sections accrued by 1793 before they were subverted by the Convention. The titles of their various *comités* and *commissions* do not tell us how intensely active the sections had become in the administration of their neighborhoods and finally in running the entire city of Paris. They were sources of information on local counterrevolutionaries and grain speculators, both of whom existed by the nestful in the city. They became the dispensers of a rough-and-ready popular justice, the guardians of the "maximum" that regulated the price of staples, the dispensers of food for the poor and sustenance for the widowed and orphaned, the caretakers of refugees from the provinces and the homeless who all but lived on the streets. They were the authentic organizers of the great *journées*, the insurrectionary "days" that pushed the revolution in an increasingly egalitarian direction much to the dismay of the Jacobins, whom they mistakenly worshipped with almost unthinking reverence before they were betrayed by the Robespierres, Dantons, and Saint-Justs. They forged the "holy pike" that armed the revolutionary populace and made it possible for it to intervene directly in the Convention's proceedings and alter the course of its decrees and laws. Going even further, the sections began to intervene significantly in the economy of the city. Their *comités d'agriculture* scoured the countryside to feed the city, their committees to establish *ateliers de charité* freely appropriated workshops abandoned by wealthy émigrés and used them to provide jobs for the unemployed who made uniforms, weapons, and gunpowder for the embattled revolutionary armies at the front. Not unlike the Athenian festivals of an earlier age that did so much to cement the *polis* together, the sections not only called periodic *fêtes décadaires* and neighborhood *banquets populaires* but helped to stage the fervent citywide celebrations that occurred throughout the revolution, fostering a sense of civic fraternity that Paris was to rarely see again after 1793, when the "year of the *sans culottes*" passed into history. And it was these *canaille*, or "rabble" as they were also called, who were to be pushed from the stage of history and shot down by the thousands in the reaction that followed the tenth of Thermidor (July 28, 1794), when Robespierre and his followers were guillotined.

Finally, "From the principle of popular sovereignty," observe Furet and his colleagues, "the *sans culottes* went as far as the notion of direct government. With the same verve which made them rise against the Girondins, in the summer of 1793 certain sections demanded to keep their own civil records, to levy their own taxes, and to render justice without appeal. No limitations should be imposed upon the assembly of citizens, which decided everything. A decision by the Convention (September 9, 1793) having limited section meetings to two days a week, the *sans culottes* got around the decree by holding a daily meeting of the 'Society of the section.' This assembly claimed to supervise the civil servants and magistrates, who in their eyes were merely their subordinates. Members of the *comités civiles*, revolutionary commissions, police commissioners, or commissioners for supplies were subjected to constant surveillance by the sections, at least as long as the latter maintained their independence from the revolutionary government. Censure, 'purifying' votes, and revocability of elected persons were the principal means by which this popular sovereignty was exercised, according to the Rousseauean principle of a sovereign and indivisible popular opinion."[22]

The well-to-do classes of France had everything to fear from these sectional assemblies and a radical Commune of Paris. Although it was highly unlikely that the largely bourgeois districts of 1789 ever envisioned a France governed by a completely popular direct democracy, the rise and growing influence of the sectional assemblies clearly evoked images of a confederal France structured around sections similar to the Parisian in all the cities and towns of the country—a nationwide Commune of communes. The extent to which this vision of an entire European country began to take hold of the popular imagination as early as November 1792, is suggested by a statement that the *Section de la Cité* circulated for the approval of other sectional assemblies. "The citizens of Paris declare ... that they recognize no sovereignty except a majority of the communes of the republic..." the statement provocatively declares, "that they only recognize deputies in the Convention as composers of a draft constitution and provisional administration of the republic." By the spring of 1793, such views ceased to be isolated. On April 29 of the following year, the *Conseil General* of the Commune, which is to say the broadest of the concentric groups within that body, took practical steps toward implementing the idea of a federation of communes by establishing a corresponding committee to communicate with the 44,000 municipalities of France. In a printed circular to the municipalities, the committee's secretary boldly declared: "That is the only kind of federalism the people of Paris want.... All the communes in France should be sisters."[23]

For all his biases and preconceptions, Daniel Guérin is on solid ground when he resolutely concludes, "The bourgeois philosophers who had pronounced direct democracy unworkable in large countries [no less Rousseau than other social theorists of his day], on the grounds that it would be materially impossible to bring all the citizens together in one meeting, were thus proved wrong. The Commune had spontaneously discovered a new form of representation more direct and more flexible than the parliamentary system and which while not perfect, for all forms of representation have their faults, reduced the disadvantages to a minimum." Indeed, it is doubtful if the Commune and the sectional assemblies could be regarded as a system of "representation." The *sans culottes* had gone much further than any assembly form of which we know in establishing a direct democracy, this in the very heart of a nation-state that the Jacobins had centralized to a degree that would have been the envy of France's most absolutist monarchs.

Chapter Six
From Politics to Statecraft

It remains one of history's great ironies that the city, which reworked stagnant archaic systems of corporate life based on status and kinship into the innovative, free realm of politics and citizenship, was to produce the very factors that led to its own undoing. European cities, I have pointed out, were different from their ancient counterparts because of their inherent autonomy as civic entities. The increasing separation of the medieval town from the city's traditional agrarian matrix produced not only a new kind of city with an identity of its own; it also produced a new type of economy, culture, and political structure that profoundly altered the countryside and slowly remade it into the city's image. Today, we have no difficulty in recognizing this profound change—the "urbanization" of the land—as a logical step in a mythic ascent of societal life from what Marx lamely called "rural idiocy" to what we like to call "civilization." Modern agribusiness is the patent "conquest" of agriculture by industry, a city-born enterprise and technics; so, too, is mass culture, which has urbanized as well as homogenized agrarian lifeways. What we do not fully sense is the extent to which early city dwellers would have regarded such a sweeping urbanization of the countryside as peculiar, nor could they have anticipated the extent to which it would have undermined civic life and citizenship. We have simply lost contact with the problem of urbanization as antithetical to citification—and to the extent that we are oblivious to the very existence of the issue itself, we have become its mute and unknowing victims.

How, then, did this remarkable change from civic autonomy to civic supremacy come about? And in what sense, institutionally and economically, did it begin to challenge the city's integrity, ultimately to raise the very real problem of its subversion as a realm of genuine politics and meaningful citizenship? Our answers to these questions, so crucial to an understanding of modern urbanization and the threat it poses to the city, oblige us to examine the new kind of economy and values that became preponderant in the communes of the late Middle Ages and the role they played in replacing civic life with the nation-state or, more precisely, politics with statecraft.

We must first look at the new economic relationships that began to link European cities and regions together—and I say "first" not because they are the sole, the "definitive," cause that produced this new economic and political dispensation. Whether the European continent "necessarily" would have been changed from a loose association of towns, cities, baronies, duchies, and the all-presiding, if ineffectual, Holy Roman Empire into a clearly articulated group of nation-states is a problem in divination, not in social analysis. How Europe could have developed—whether toward confederal communities or toward highly centralized nation-states—is an open question. One can single out many reasonable alternatives European towns and cities might have followed that were no less possible than the one that became prevalent in fairly recent times. No single course of development was "inevitable" or "predetermined" by the economic, social, and political forces at work. Indeed, that Italy did not become a nation-state until the nineteenth century must remain an utter mystery if a constellation of cultural, social, and political factors, particularly the role of its cities, is not invoked against strictly economic explanations. Seemingly, no area of Europe was more "modern," "capitalistic," or entangled in a market economy some 500 years ago than the Italian peninsula, and yet Italy was to lag behind western Europe's trend toward nation-state building for centuries. Nor can we understand why it was only in England that a market economy virtually absorbed all other economic forms of life such that the British Isles became the "model" for a capitalistic society in the nineteenth century while Spain, which entered so early on in the development of nation-states, lagged behind Europe as a whole and remained a predominantly agrarian society until well into the 1930s.

This much is clear: from the thirteenth or fourteenth centuries onward, Europe—and, most notably, Italy—was the scene for an entirely new economic and social dispensation. The Italian city-states began to break with traditional economic relationships that had been ingrained in ancient and early medieval civic life ways. From Italian ports and inland commercial cities—and in northern Europe from Flemish industrial centers—foreign trade began to bring a growing number of parochial communes into a new kind of commercial network, one that resembled what we would now regard as capitalistic. Trade, accumulation, and the reinvestment of profits into expanding and competitive business enterprises now became ends in themselves, not simply means to achieve personal wealth, land holdings, and aristocratic status. Not that this development was totally universal, as we shall see very shortly. But it occurred on a sufficiently wide scale to make it unique and to subvert the traditional agrarian lifeways that marked European society, merchant as well as noble, in past historical eras.

Yet new as this development was, I do not wish to overstate it. Our own society, a society that celebrates the ascendancy of "free trade" and a competitive capitalist economy, is ideologically imperialistic: it tends to cast the past too much in its own image. All the roads of European history do not lead to the triumph of the twentieth-century market economy. The fact is that feudal values, rooted in an elaborate system of rank and a strong orientation toward the ownership of land, were no less a part of received wisdom of the new, rising urban "bourgeoisie" than of the old agrarian nobility. Trade did not alter this received wisdom of feudal society for centuries to come. Indeed, like the merchants of antiquity, whose goal consisted of amassing enough wealth from the "sordid" operations of commerce in order to retire to a manorial life in the countryside, many Italian and Flemish merchants had very similar ambitions. Prestige gained by titles through intermarriage and by the ownership of landed property was still a desideratum among these early "bourgeois"—and such goals were to remain widespread in Europe well into the eighteenth century. On this score, the merchants of antiquity and their descendants in the Middle Ages were very much alike in outlook, particularly the most wealthy ones. In their psychology and sense of cultural self-definition, the city rich were similar to all the upper rural strata of the time. And we shall also see that these ambitions and this mentality exercised less of a hold among the more middling sort of city dwellers, such as artisans, ordinary merchants, and aspiring members of the plebeian orders. Differences in attitudes and wealth were to profoundly divide the medieval towns and cities internally, greatly complicating their relationships with the emerging monarchies.

For the present, it is important to emphasize that the wealthy merchants of the late Middle Ages differed from their ancient counterparts in the *way* in which they were committed to commercial operations. Although their outlook closely resembled a feudal one, their practice was very akin to a modern one. The tension between the old and new, between precept and practice, introduced ways of functioning that deeply altered European life. Ancient trade, generally speaking, was surprisingly simple in comparison to later medieval business transactions despite its Mediterranean-wide scope. Ordinarily it was local and more like barter than modern forms of exchange; the ancient economy was not as highly monetized as it is today. Money was conspicuous more by its absence than by its presence, and the extent to which ancient trade was a regulated affair would have chilled modern acolytes of *laissez-faire* doctrines.

Like the cities, which were usually religious and administrative centers, often as parasitic in their need for tribute from the surrounding countryside as in the exactations they placed on distant subject peoples, Mediterranean commerce found itself physically and culturally hemmed in by a distinctly agrarian world and a largely subsistence economy. Beyond the ancient cities that clustered along the shores of the Mediterranean or were planted strategically at intersections of inland waterways, the early trader faced a semi-feudal world of crude manors and impoverished peasant villages, a world that dissolved into a forested, semitribal communal world. Both of these worlds, the manorial and the tribal, in antiquity constituted a very precarious, indeed hazardous, terrain for the merchant, for his caravans on the land and his ships in remote waterways.

The ancient merchant responded to these barriers accordingly. Trade was distinguished by its highly personalized and somewhat tentative character. The merchant and his sea captain or caravan leader were united by a specific enterprise rather than a highly organized business, although there were always many notable exceptions to this rule. Ships often went to sea and caravans with pack animals went inland to make a "killing" just as a wayfaring stock-market speculator today seeks his "lucky break." In this sense, merchants were literally "merchant-adventurers," and their "companies" had quasimilitary characteristics. They commonly went abroad as armed expeditions. Loans were often no less personal than the expeditionary techniques that brought a fleet of ships or a caravan of pack animals together. Although credit was fairly well developed by Roman times when the empire was already quite secure, in foreign trade to fairly remote parts, at least, credit was often seen more as a gamble than a safe investment. Such profits were high, and went into hoards or were invested in land rather than the expansion of on-going businesses. I speak here of foreign trade, not local trade; of the ancient world's "merchant-adventurers," not its home-based business enterprises, artisans, and commercial farmers.

What the Italian merchants introduced in the Middle Ages was as close to a "revolution" as anything that goes by that name. A broad network of business houses, credit institutions, trading depots, warehouses, and affiliates began to interlink Italian city-states with numerous medieval European towns, and the movement of goods was gradually secured by treaties, tribute, and hired mercenaries from the predations of robber barons and bandits. The Venetian navy virtually swept the city's Mediterranean trade routes of pirates, attaining a sea supremacy that rivaled the naval power of the more formidable empires around the basin. Business became systematic, safe, a predictable enterprise. Money was invested not only in the expansion and acquisition of landed property; it was also invested in larger commercial operations whose profits not only attracted ordinary people but also the local nobility. The lure of this fairly safe source of wealth, still disdained culturally but crassly attractive to an increasingly practical and secularized world, was enormous. It opened new doors everywhere in late medieval society. Indeed, the social mobility of the fourteenth and fifteenth centuries sharply contrasts with the social stagnation of the classical world. Well-to-do commoners were closely interlinked with nobles by marriage as well as by trade. Aristocrats offered titles and pedigree in exchange for the wealth and material security offered by merchants. Alliances were forged not only with families of the same social order but between families of different social orders. This melding of "classes," largely feudal and "bourgeois," has only recently received the attention it deserves and was to have a profound effect on the so-called "bourgeois revolutions" that marked the eighteenth century.

Finally, where elite orders were not interlinked by marriage, merchants and even professionals acquired aristocratic status by buying it outright, especially from the fifteenth century onward. A considerable portion of the revenue acquired by the French Bourbon kings, for example, came from the sale of titles, a sale that united the higher orders of society by a shared sense of titular nobility—even as it divided them by a sense of shared disdain between the purchased "nobility of the robe" and the hereditary "nobility of the sword." In time, alliances and differences were to become as tentative and precarious as distinctions formed by wealth, when the rich of one day could easily suffer the pangs of penury after the misfortunes of another day.

What did last and expand, however, were the trade networks established between communities, particularly the larger towns of medieval Europe. But the enormous growth, continuity, and increased stability of this foreign trade—the nearest thing to a capitalistic enterprise the continent developed—should not cause us to overlook the rich elaboration of the local market that occurred within the medieval communes and their immediate environs. The decline of the Roman Empire and the eclipse of the administrative cities that held it together brought the European continent face-to-face with itself and threw it back on its own mainsprings.

No longer could the continent be sustained and civilized, much less managed and exploited, by the Mediterranean basin with its rich granaries and sources of tribute. Beyond the highly cultivated Roman-controlled territories of Europe lay great forests and bogs, a huge continent within the continent itself that was fairly pristine and open to new forms of social development. Communes began to appear in this virginal terrain that were now structured not only around religious and governmental institutions but, significantly, around the markets of craftsmen and artisans. These artisan towns, to be sure, were crude and unpolished; they were more villagelike than urban. The work and the markets that emerged from this highly decentralized, localistic society were still mainly organized around religious and feudal values. Often collecting around churches and cathedrals—indeed, commonly under the sovereignty of a bishop— they nevertheless became productive towns with small markets and family-operated workshops, living off the agricultural produce in their environs and providing rural folk with the more skillfully crafted commodities they could not make in the countryside.

Here we encounter something remarkably new: communes marked by a rich social life, and with it a popular politics rooted in guilds, systems of mutual aid, a civic militia, and a strong sense of community loyalty. The guild, the most important institutional center of a new kind of artisan society, should not be confused with the ancient collegia that it resembled structurally or with such strictly economic associations as modern trade unions, which it seems to resemble functionally. The Roman Empire exhibited a very low tolerance for nonstatist bodies that could challenge the authority of the monarchy and its bureaucracy. Hence ancient collegial associations of craftspeople were rarely permitted to extend their activities beyond those of burial societies and festive fraternities. They were allowed to have no economically regulative or protective functions.

Essentially, they were cultural and quasireligious institutions. Modern labor unions, in turn, are primarily economic organizations and assume few cultural, much less religious, undertakings.

The medieval guild, by contrast, assumed the responsibilities of the ancient collegium, the modern labor union, and a great deal more. Possibly religious in origin, it became highly secular in many of its activities, fostering material as well as moral commitments. It was a sworn, covenanted brotherhood that punished its members for lowering the quality of goods or seeking higher than prescribed prices. It regulated personal, moral, and religious behavior as well as the output of goods. Not only did it care for widows and orphans, the ill and infirm of its own members, but it gave alms to the poor, performed charitable works of all kinds, celebrated feast days, punished its members for usury, blasphemy, gambling, and other presumably "immoral" infractions of good behavior.

What made medieval guilds particularly significant in ways that mark a sharp departure from towns of the past is that they attained a degree of legislative and governing authority that made them the principal municipal institution of many communes. However parochial they seem to us today, European towns by the thousands achieved a degree of autonomy that few municipal entities had acquired in times past or were to acquire later. This autonomy was pieced together corporately from a localized world of small artisans, craftsmen, and merchants—a feudal world, in fact, not a capitalistic one, however much it was centered around the marketplace—before European communes were to be networked together by foreign trade. The commune's growth and elaboration took place organically, not artificially, within a highly decentralized agrarian society. In contrast to the Roman-controlled town, the European town was a unique phenomenon insofar as local autonomy became the rule rather than the exception. Control from below thrived at the expense of an institutionally weak feudal society that was beleaguered by overlapping and conflicting jurisdictions between landed nobles, urban bishoprics, papal legates, and insecure monarchs. The unending series of conflicts that these jurisdictions generated was exacerbated by an undeveloped system of communications and by a crude armamentorium, one that made civic militias a powerful military force in social life.

Foreign trade—more precisely the capitalistic carrying trade that emanated from the Italian city-states and the Flemish communes—definitely worked against this communal autonomy as surely as it played a role in interlocking Europe's towns and cities. And in so doing, it posed a historic problem: would the commercial network created by intercity trade yield the formation of nation-states, centralized by monarchical and later republican systems of power, or would it give rise to confederal institutions, united in a shared continental system based on local community control? That the nation-state was to gain ascendancy over confederal systems of self-governance does not mean that its victory was predetermined by Europe's history—nor does it mean that its victory cannot be undone. How that nation-state's ascendancy was achieved is a story worth telling because it also brings to light an achievement that was often very tenuous, owing to the resistance of the towns to centralism, a struggle that has not been definitively foreclosed by urban issues that are emerging today.

For the present, we shall confine our account of the nation-state to the way it was achieved, with the clear reservation that other alternatives continually existed, indeed dramatic alternatives that we will explore later. Initially, the intercity carrying trade that began to unite Europe economically from the twelfth century onward not only tore into a complex web of mutual personal and communal dependencies in which trade as well as behavior was carefully regulated; it also became the infrastructure for a new body of societal institutions—initially regional rather than local, later overwhelmingly national—that cut across the grain of a time-hallowed and intensely communal and decentralized mode of social life. It eventually introduced nationalism, a distinctly European phenomenon that was to spread beyond the continent itself and acquire global dimensions.

The ancients, like the early Europeans, had very little experience with the notion of nationhood. Largely tribalistic or localistic in outlook, they tended to look inward toward their traditional lifeways, to elaborate them rather than innovate new institutions and values.

Even the Greeks and Romans, who were comparatively "forward looking" in their attitudes, were heavily guided by tradition. Cultural as well as economic "limits to growth" were deeply molecular: people owed their strongest allegiances to their kin group; next, to their community or perhaps region; rarely to a "nation." The idea of a "nation" was alien to the ancient mind, a tribalistic form of mind that opposed the locality to the ecumene.

Although pan-Hellenism was very much in the air among the Greek *polei* shortly before Alexander brought the western world and the Near Eastern together institutionally, it quickly drifted into a cosmopolitan Hellenistic ecumene that adopted Greek for its *lingua franca* and Greek culture for its spiritual adornment. Hence a Greek "nation" never developed among the Greek *polei*.

Israel seems to have acquired a strong sense of nationhood after the Maccabean Revolt, but it was smothered by foreign invaders. Religion ultimately placed a stronger claim on the Jews than a sense of territorial nationhood, hence a budding form of nationalism was soon supplanted by a powerful belief in a spiritual community whose strength still defies the economistic explanations advanced by crude variants of Marx's "historical materialism." The great empires of the ancient world were not "nations" in any sense of the term. Indeed, it is difficult to associate them with the modern, class-based nation-state. In the Near East, these empires assumed a highly patronymic form: a "property," as it were, of a deified, patriarchal monarch for whom the vast lands under his control were regarded as part of his *oikos* or household. Lands annexed to the monarch's original inheritance became the tributaries of a centralized household rather than the territory of an institutionalized body politic. The Roman Empire, particularly in its imperial rather than republican form, inherited this Near Eastern tradition-bound state form, however much it was secularized and regulated by laws. Indeed it should be kept in mind always that the Roman state was managed primarily by patricians, a group of "fathers," by definition, rather than by citizens. The emperors, in turn, were the "fathers" of their people, not simply their sovereigns,

European nations, by contrast, were pieced together by sterner stuff. However much monarchical nation building, so redolent of ancient statecraft, went hand in hand with the market's expansion, the infrastructure created by commerce laid a stronger foundation for nationalism than anything we encounter in the ancient world. This new, continent-wide commercial nexus, formed out of the interlinking of towns, produced material dependencies for goods that cut across the moral relationships fostered by traditional society. Even before the monarchs of Spain, France, and England asserted their authority over their respective nations, they shrewdly exploited the divided loyalties and value systems created by a growing commercial dependence on far-flung markets on the one hand and a powerful psychological dependence produced by a richly elaborated localist community on the other. Between the material wealth offered by the former and the spiritual security offered by the latter, many towns were to divide internally as sharply as they divided against the countryside. Hence the European commune was pulled in two opposing directions: between the desirability of the nation-state and an ideal of communal confederation. France was to provide an existential example of the first, the Swiss Confederation, in its early days, of the second.

Such alternatives did not really exist in the ancient world apart from Greece. The administrative cities of antiquity easily dissolved in the West into virtual villages with the decline of the empire, villages that formed the real community base of that society. Indeed, once sources of tribute disappeared, the resources for maintaining the Roman imperial state disappeared as well and the West devolved into decentralized feudal society. Europe fell back on its own resources and its own authentic forms of social and economic organization.

The long history of Europe's medieval development, a development that brings us to the opening of the modern era, totally changed the setting for urban evolution. European-wide trade, centered entirely around a new kind of self-sustaining municipality, opened sharply contrasting opportunities for development. The wealthy elites of the towns were riddled by the divided loyalties and interests to which I have already alluded. Among the rich merchants and their noble urban allies, a growing trend surfaced for unfettered trade, free of guild restrictions and traditional moral constraints.

On the other hand, the great majority of artisans, journeymen, small retailers, and professionals were to demand the perpetuation of traditional controls, their time-honored source of security and stability. What complicates this fairly conventional account of the "class struggle" within the medieval city is the very disconcerting fact that the medieval patricians and plebeians often united as readily as they fought with each other, notably against assaults from outside the city itself. Although by no means consistently, both "classes" commonly joined together to support their municipal privileges against landed nobles and, more strikingly, even against monarchs who were bent on achieving a highly centralized state. While the ordinary plebeian strata tended to be more consistent in their commitment to their civic rights, the more conflicted patrician stratum—feudal in outlook but decidedly bourgeois in its commercial practices—oscillated in its loyalties between the more popular elements in the city and the elite elements, noble and kingly, to which it felt a groveling loyalty that marks all parvenu elements in society. As we shall see, such alliances within the commune tended to be tentative and inconclusive, at times giving rise to serious urban revolts against the newly emerging absolute monarchs, at other times dividing the towns so seriously that they easily fell prey to monarchs who were to fashion the modern nation-state.

The reader who looks for a compact development toward a modern urban society will not find one here. Between the fourteenth and sixteenth centuries, when nation-states began to form, the continent's towns and cities found themselves deeply entangled in a skein of shifting alliances and conflicts. Modern notions of "free trade," largely confined to merchants engaged in the carrying trade of the Mediterranean basin and the continent, were nevertheless permeated by feudal values rooted in land ownership and status to an extent that would have seemed curious to the modern mind. "Capital," in effect, was at war with itself and had very little identity of its own. Newly emerging capitalists of the high and late Middle Ages were often pulled in sublimely contrary directions by a fading past and a barely emerging future.

For their part, the plebeian elements of the towns were deeply committed to their ancient traditions, that is, to the corporative values of the medieval world. No affinity to "free trade" or notions of unregulated business practices were to be found among these less privileged groups. Despite their quasifeudal notions, however, they resolutely opposed territorial lords who sought to challenge the liberties of their towns, ironically in contrast to the wealthy merchant strata in their own communities who feared both camps in this conflict and variously allied themselves with one against the other. Finally, the newly emerging princes and monarchs who were to eventually piece together the nations of Europe were themselves deeply divided by this skein of tentative alliances and conflicts. They were more than willing to use the European communes against nobles who challenged monarchical authority. But they were equally willing to subdue the communes when they raised the cry for their ancient municipal liberties and autonomy.

It is within this highly unstable yet very traditional world—unprecedented in the ancient Mediterranean—that a new form of societal organization was to permeate very old ones: a centralized state apparatus structured around a distinct national entity. The ancient world had seen the centralized state in all its grandeur and power. And it had even seen in a bare, rudimentary form the outlines of the nation. But never before had the two—state and nation—been cojoined to produce a form of statism based on nationalism with its far-reaching sequelae for the modern world and the emergence of a highly corrosive global market economy.

Within this rich, highly variegated, and fluid period of history, when societal development could have followed very different directions from the one toward which it moved, we must reexamine the conventional wisdom about the rise of the modern state and the formation of the nation.

That the state preceded the nation historically hardly requires emphasis here. The previous pages of this work have operated with this fact throughout. The professional institutionalization of power and the monopolization of violence by distinct administrative, judicial, military, and police agencies occurred fairly early in history. The state, so conceived, emerges as a highly compact entity whose persistence from ancient times to the present seems almost unchanged functionally, however much it has varied in form.

But so functionally similar an institution, seen only in terms of its "class character" and its coercive role, can be very tricky when it is treated without respect for the many nuances of its development. Among many social theorists, this simplistic notion of the state has given rise to images of various state forms as mere epiphenomenal expressions of a basic, deceptively unchanged structure—indeed, to use Marxian language, as the mere "superstructure" of an unaltered "base," hence an institution that scarcely deserves searching analysis. In modern politics, this simplistic approach produced considerable mischief. The careless use of words similar to "fascist," applied to established republican states as well as totalitarian ones, can generate very slovenly and crude political attitudes. States have been called fascist or, for that matter, democratic that are very far from being either one or the other; and their opponents have often disregarded hard-won civil liberties that deserve the most earnest support by any ideological standards or values.

No less disturbing, there has been a gross disregard of democratic rights generally among self-styled "progressives" whose concerns for material justice have supplanted their concerns for social justice. The primacy given to economics, an emphasis uniquely characteristic of a market-economy mentality—and most evident, ironically, in socialist and syndicalist ideologies—has led to a troubling disregard for libertarian political institutions whose preservation and expansion is of immeasurable importance to the development of a new, municipally oriented politics. These institutions have often been contemptuously dismissed as "bourgeois" by many socialists and anarchists alike, although we shall have occasion to see that the "bourgeoisie" was never libertarian in outlook and rarely republican in its commitment to state institutions. Even liberal and conservative ideologies have used such words as "freedom" so ecumenically that both the content and form of the term have been absorbed into a meaningless "black hole" of sociological rhetoric.

The interchangeable use of the words "state" and "nation," in turn, has been even more troubling. Trite avowals of "my country right or wrong!" imply no commitment to a republican state, even when the cry is justified by a country's seemingly democratic or libertarian institutions. Patriotism today is nationalism, not democratism, although during the French revolution the two words could have easily passed as synonyms. "Love of country" may characterize the sentiments of a fascist, socialist, liberal, or conservative, and this love does not in itself commit any one of them to a particular state form, much less to a free community. Hence the development of the state and the emergence of the nation are not matters of academic interest. Their history is deeply entangled with the prevalent societal values of our time and profoundly affects our visions of society's future, especially in a discussion of the municipality.

A close study of the state shows that there are and have been varying *degrees* of statehood, not simply the emergence of a finished phenomenon called "*the* state." Indeed, the universal use of such words as "state" can impede a clear understanding of the extent to which "the state" exists at various levels of societal development—not only historically, but also today in modern society. Conceived in a processual way with due regard to the degrees of statism that have existed historically and functionally, I should emphasize very decidedly that "the state" can be less pronounced as a constellation of institutions at the municipal level, more pronounced at the provincial or regional level, and most pronounced at the national level. These are not trifling distinctions. We cannot ignore them without grossly simplifying politics. Differences in degrees of statification can have major practical consequences for politically concerned individuals and communities.

History, moreover, provides us with compelling evidence of germinal states, quasistates, partially formed, often very unstable states, and finally fully formed and all-embracing states. The Athenian *polis* and even the Roman Republic were not fully formed states—this in contrast, for example, to the fairly well-formed Roman imperial state and even more fully formed Egyptian state of Ptolemaic times. When applied to classical Athens, the use of the word "*state*" has a very limited meaning, despite the presence of slavery. Even in the modern nation-states we know best, municipalities are often less "statified" than nations, with the exception, to be sure, of the patently totalitarian states that have emerged in our own time. In practical terms, a modern municipal politics can be very different from a national parliamentary "politics," as we shall see in the closing chapter of this book, and localist politics can rest on ideological traditions and premises very different from those we associate with the formation of the nation-state.

The fact is that we have been much too concerned with the origins of the state, conceived primarily as an instrument of class domination, to give due recognition to the history of the state: its evolution, its various unfinished forms, its varying kinds of structure, and its capacity to penetrate the social and political life of the community as well as the nation.

Living as we do in "founded" republican states of one kind or another—states that are clothed in a panoply of "declarations," "constitutions," "charters," and even highly personified "fathers" and "founders"—our images of "the state" tend to acquire a highly contractual, legalistic, and contrived form. "The state," with its clearly dated documents, seems more like a social contract than a historically conditioned phenomenon. Behind the "contractual" state lies an anthropology and history that, carefully considered, desanctifies its rationalistic claims to authority and its mandate as the source of an orderly society. The state, in fact, had to fight its way into existence against claims that were no less rationalistic and morally valid than those that it advanced on behalf of its own legitimation. It had to emerge organically, that is to say, within the framework of social relationships and, later, political norms that were by no means consistent with and were, at times, highly antithetical to the formation of a state apparatus. Hence, it is fair to say that just as the constituted or constitutional state preceded the formation of the nation, so an organic state, uncertain of its pedigree and of dubious legitimacy, preceded the establishment of a constituted or "constitutional" state. The organic origins of the state, in turn, bring into question the extent to which the state can be validated wholesale on strictly rationalistic terms and, above all, its capacity to absorb the very aspects of social and community life in which it was gestated. To demystify the authority of the state as a rationalistic contrivance is to take the first step toward recovering the Hellenic notion of politics as a public activity, the domain of authentic citizenship—not as statecraft, the domain of the professional legislatures, military, and bureaucrats.

Contrary to rationalistic and contractual images of the state, state institutions emerged slowly, uncertainly, and precariously out of a social milieu that was distinctly nonstatist in character. In fact, the social and organic sources of the state had to be meticulously reworked before they could give rise to state institutions. The ancient temple corporation, actually a religious legitimation of tribal collectivity and public control of land, seems to have been the most likely source of the Near Eastern state. This was a time when priests commonly became kings or, at least, when the kingship often took on a priestly character. In either case, the temple and palace monumentalized as well as deified the tribal community.

Despite the increasing secularization of the state, notably in Greece and Rome, the state never completely lost its religious trappings and its function as the custodian of the collectivistic community. This attribute, whether as an ensemble of feudal nobles or a monarchy and ultimately as an absolutist empire, remained with it well into recent times. The traditional "head of state," be he a lord or king, always remained the "father of his people," whether by divine right or as a divinity in his own right. Hence, prior to the rise of republican systems of governance, the state always appeared not as a constituted phenomenon but as a reworking of a very traditional, organic, patriarchal, indeed tribalistic body of relationships in which power was not simply conferred by the community as in the case of elected kingships but inherited along lineage and blood lines in a manner redolent of the ancient tribalist blood tie. The present always entailed a reworking of the past, a transmutation rather than a dissolution of traditional forms to meet new needs and imperatives.

It is notable that the rise of the centralized nation-state in Europe also followed this archaic and highly organic process of transmutation of old into new. Indeed, until "The Age of the Democratic Revolutions," to use the title of R. R. Palmer's distinguished book, it was not through the constitution of new states but the recovery of ancient rights that king and community were thrown into civil war with each other, a conflict that often took the shape of monarchy against municipality.[1] Neither one party nor the other sought to innovate new forms of governance but rather to restore old ones from the past. Characteristically, the earliest form of the European nation-state appears not as the emergence of a national economy, significant as this development proved to be, but as the increasing sovereignty of the kingly household itself—the monarchical *oikos*—and the image of the "nation" as a kingly patrimony.

The evolution of the kingly household into an authentic state is strikingly revealed by the evolution of the English monarchy. The reputation of England as a uniquely centralized state from the days of the Norman Conquest in 1066 tends to be overstated. Admittedly, William the Conquerer took firm possession of Anglo-Saxon England shortly after the defeat of Harold in the Battle of Hastings, but the area under Norman control was relatively small, almost provincial in size. Wales and Ireland had yet to be conquered and Scotland to be absorbed. William's absolutism was not only restricted territorially; it was short-lived historically. The English state—and certainly its highly fragmented sense of nationhood—is notable not for its continuity but its discontinuity. Growing baronial strength clearly began to abridge monarchical rule in little more than a century after the conquest: the Magna Carta, to which John unwillingly set his seal at Runnymede in June 1215 is testimony not to the rise of English democracy, all legends about the charter's intent aside, but to the power that the English barons acquired at the expense of the monarchy. Although John's father, Henry II, had left his sons a state buttressed by a system of royal law remarkable for its time—extending the "King's peace" to include civil and criminal cases, a rationalized system of trials, punishments, and juries, and a professional royal judiciary to translate this system into practice—many of these jurisdictions were to be reclaimed by the barons. Nominally centralized, England remained remarkably decentralized under the weaker monarchs who filled the long span between Henry II and Henry VII, a period of some three centuries. A centralized infrastructure had emerged from the conquest, but history had yet to flesh it out with effective royal institutions.

What makes the English state interesting is the challenge it raises to simplistic theories of state formation and rule. I refer to its organic roots and its evolution out of household offices. The English state was born not out of an administrative body of autonomous departments but rather it was formed out of the personal responsibilities of the king's servants—his immediate household coterie— often in opposition to the doubtful loyalties of the king's own feudal barons. Perhaps the foremost of these royal servants was the king's personal secretary, his chancellor, who carried the royal seal and coordinated the emerging departments that comprised the administrative portion of the royal court. In time, the chancellor became the pole around which an increasing number of clerks, experts, and specialists in various governmental areas, and overseers of what was to become a fairly complex executive authority collected to form the all-important English chancery. Almost every aspect of monarchical rule fell within its purview, principally the king's exchequer who saw to the collection of taxes and Henry II's professional judiciary.

In fact, the English state was formed largely from the king's bedroom, dining table, men-in-waiting, and household clergy, not from constituted principles of government that spoke in the interests of a specific "ruling class." Class theories of the "origins of the state" to the contrary notwithstanding, the English state of the Middle Ages began as the elaboration of a patrimony rather than the institutionalization of one "class's" authority over that of another. The English barons, who were to view the formation of this state with suspicion and later with overt hostility, found it difficult to claim it as their own. A continual tension existed—occasionally expressing itself in a violent form—between the baronial infrastructure of English medieval society and the monarchy, which formed the originating impulse of an authentic, fairly complete state. In its patrimonial form, the English state is no exception to the "origins of the state" generally; this mode of state formation is very similar to the way in which the "barbarian" chiefdoms of an earlier tribal society gradually extended their power from networks furnished by their personal retainers and clans. The journey from "valet" to "prime minister," amusing as the juxtaposition may seem, is closer to the truth of state formation than the more "sociological" idea that the state emerged as an agency of class interest—whatever it was to become later in history.

I have dwelt in some detail on the origins of the English state—in time to be regarded as the prototype of the nation-state *par excellence*—not because of its uniqueness but rather because of its continuity with the ancient past. The organic growth of the English monarchy parallels to a remarkable degree the rise of *oikos* forms of statehood. Historically, these forms go back to early Egypt, Persia, Babylonia, and even Rome before the empire became heavily bureaucratized.

Why did the English state become such a useful framework for the modern nation and for modern capitalism? An answer to this question lies precisely in the limits as well as the rationalistic and centralistic forms it assumed, both territorially and institutionally. Despite its exhausting adventures in France, the English state by Tudor times remained focused more fixedly on its own island territories than did its archaic antecedents. And institutionally, it never achieved a degree of absoluteness—or at least was never permitted to do so—such that it devoured its own bourgeois "golden goose." It permitted a flexibility of market development that was rarely to be seen in the past, not only in the medieval world. Thus, while the absolute monarchs of England (principally Henry II and the Tudors) managed to hold a nation together, at least in the Anglo-Saxon core areas of the island, eventually integrating the Scots as well as the Welsh within a national framework, it provided ample space for its middle-class "commons" to flourish, prosper, and, in time, resist the exactations and arbitrary demands Charles I imposed upon them. Throughout much of its history, England was burdened less by a bureaucracy than her rivals in western Europe. While the monarchy retained a well-knit bureaucratic structure, particularly in the administration of royal law and the collection of taxes, England was largely managed by its local squirearchy, a highly personalized system of management guided by the rationalized standards introduced by Henry II and the Tudor monarchs. The English Revolution of Cromwell's time finalized this delicate balance between the king and his "commons" in a way that imparted a fictive quality to the English state as an ideal "bourgeois" political system. Indeed, it was England's "constitutional monarchy" that was to have an almost hypnotic attraction for the progressive intelligentsia of the French and German enlightenments.

Ironically, it was France rather than England that was to create the kind of all-pervasive bureaucratic nation-state that characterizes present-day bourgeois state forms. Given the waywardness of history that defies attempts by historians to systematize the development of events in the name of historical materialism and other "scientific" explanations of the nation-state's emergence, France's preeminence as the prototypic nation-state is explained not by her "advanced" development as a "bourgeois" society but rather by the lateness of that development, indeed by delays that were to significantly initiate that development in the mid-nineteenth century, well after it occurred in England. To be sure, French absolutism did not emerge in a sudden burst of centralization but quite to the contrary. Louis IX ("Saint Louis"), more than a century after Henry II of England, still issued his decrees with a terminology that is redolent of the early Frankish system of collective rule in which the Germanic kings were considered merely first among equals. The expression "We and our barons" recurs in Louis's pronouncements, a phrase that in no way suggests a commanding authority over the feudal community. But even more so than England, France began to enter into a phase of state evolution that was to induce many historians to regard French absolutism as the predecessor of all Europe—an overstatement, to be sure, but a sufficiently suggestive one to impart to the French Revolution, which overthrew it, a particularly incendiary challenge to absolutism as such.

By the end of the twelfth century, France had already begun to catch up with England by creating *les officiers du roi* (officials of the king) who shared power with the French barons in the traditional royal council. By degrees, the French began to outpace their English rivals. Functionaries, emerging from the royal household, acquired expanding administrative roles so that the kingly servants were soon to be royal bureaucrats rather than household administrators. In contrast to the English monarchy, the French carried this development much further: it encompassed time-honored *local* as well as royal jurisdictions. Already a huge hierarchy of petty officials had arisen, such that by the end of the thirteenth century, Philippe le Bel was obliged to place the host of *lieutenants, sergents,* and *bedeaux* who afflicted the French people on local and provincial levels under the scrutiny of *controlleurs,* a royal strategy that may have enhanced rather than diminished the bureaucratization of the nation. Royal *commissaires* were to become permanent regional officials by the mid-sixteenth century and a far-reaching network of *intendants,* supervised by *surintendants,* acquired the odious status of a financial bureaucracy that particularly aroused popular hatred.

In time, the immense French bureaucracy of the sixteenth and seventeenth centuries, in theory answerable only to the monarchy, acquired a life—indeed an outlook—of its own. The emergence of a bureaucratic sensibility, permeating all levels of French society, can hardly be emphasized too strongly. In contrast to so much of feudal Europe, the sons of the French middle classes began to regard the royal bureaucracy rather than the clerical hierarchy as the avenue toward upward mobility and power, a shift in perspective that linked the French "bourgeoisie," whatever that word meant some two centuries ago, to the monarchy more tightly than historians of "class conflict" would have us believe. The French Revolution, conceived as the "classic bourgeois revolution" of emerging capitalism, was to test this "class analysis" in the fiery crucible of insurrection, with more dismal results than later, nineteenth-century historians suspected.

Herder's conclusion to the contrary notwithstanding, France was by no means the "precursor of all Europe." It is beguiling to think that an even more centralized and bureaucratized "nation-state" was established in Sicily in the early thirteenth century, when Norman conquerors dispossessed the Arab rulers of their control over the island and established what many historians have variously called the "earliest modern state" and "absolute monarchy" in Europe's history.

Emperor Frederick II did, in fact, create an "omnipotent royal power" that led to the "complete destruction of the feudal state" to use Jacob Burckhardt's words, a state marked by a completely centralized legal system, a professional army (which the French monarchy introduced very early in its evolution), and an all-encompassing bureaucracy of professionally trained officials, all indubitable traits of Norman Sicily.

But these are traits of a kind that in no way made Sicily a "modern state," much less a "nation-state."[2] Such state forms, in fact, were to appear very early in human history. The Ptolemaic state that followed Alexander's conquest of Egypt in ancient times did not differ in fundamentals from the structure that Frederick imposed on Sicily some fifteen centuries later. Characteristically, both Alexander's general, Ptolemy, and the Norman monarch, Frederick, wedded the economy of the Nile and of the island to the state itself. In both cases, key commercial operations, particularly the grain trade, became state monopolies and economic activity was bound to statecraft. The Norman state in Sicily was an "Oriental despotism," to use the language of nineteenth- and early twentieth-century historiography, not a "modern state," much less a "nation-state." Administered by conquerors and their bureaucracies with highly regulated economies, the Norman state was no more "modern" than the Inca state in Peru or the Egyptian state in North Africa.

We are faced with the paradox that one of the earliest approximations to a centralized "nation-state" emerged in Spain, even before the Tudors in England and the Bourbons in France fully consolidated their rule. Yet Spanish absolutism and nationalism did not promote the development of a "bourgeoisie" or a "bourgeois society"—developments that have been associated with the emergence of the nation-state. In fact, Charles V and his successors were to virtually devour a flourishing Spanish burgher stratum, milking it of its wealth, preempting its power, and essentially subverting town life in the Iberian peninsula. At the beginning of the sixteenth century, a swath of almost meaningless royal parasitism and domestic extortion cut across Spanish history, eventually undermining a flourishing city network that enjoyed extraordinary wealth and autonomy. The extortions initiated by the Spanish state led, as we shall see shortly, to one of the greatest urban uprisings in western history, an uprising that decided the future of the new nation that had emerged from the Christian reconquest of the Iberian peninsula after centuries of Moorish rule. City life and commerce were to stagnate or decline for reasons that had much more to do with absolutism and its efforts to forge Spain into a nation-state than did the decentralization that marked the ensuing history of the country.

What is most intriguing is that neither absolutism nor the rise of a nation-state provides us with an adequate explanation for the rise of a "national economy" as Hannah Arendt suggests. Although Spain was to remain a largely agrarian society up to the 1930s, indeed a very traditional one, its Hapsburg kings were no less "enlightened" than the Bourbon kings of France and, with the exception of the two Henrys and Elizabeth, the Tudor and Stuart kings of England. Nevertheless, Spain moved into a period of economic decline that still weighs on her shoulders today while England became the "factory of the world" and France its cultural "perfumery." Although European nation-states from the sixteenth century onward created the arena for a national economy, they did not necessarily create the forces that shaped it. Absolutism, which, sculpted a sense of nationhood out of feudal parochialism, played a very crucial role: it not only supplanted localism with nationalism; it also stifled a highly decentralistic, localistic, and spontaneous society, marked by a rich diversity of cultural, economic, and communal attributes, replacing it with increasingly homogenized lifeways, bureaucratized institutions, and centralized state forms. In some cases, this absolutist alternative favored the later expansion of a market economy; in others, it led to state parasitism and outright regression. In all cases, however, it turned localist politics into nationalist statecraft, divesting citizenship of its classical attributes and turning vital, empowered, and strongly etched men and women into passive, disempowered, and obedient "subjects."

This shift from a living people to deadened subjects did not occur without furious resistance. A belief in autonomy, regional and local identity, and citizen empowerment ran very high between the late Middle Ages and fairly recent times. The battle to retain these distinctly political qualities and rights was to be fought not in national political parties or by professional statesmen; rather, it was conducted on the level of village, town, neighborhood, and city life, where the ideals of confederation were to be opposed to demands for a nation-state and the values of decentralization were to be opposed to those of centralization. What lay in the balance was not only the future of the town and countryside but the development of political institutions as opposed to state institutions—and an active citizenry as opposed to a passive "constituency."

It has become somewhat conventional in urban historiography to treat "city-states" as though their relationships with each other were normally marked by endless petty squabbles and their relationship to absolutism by an almost unqualified degree of support. "City-states," we are commonly told, were almost innately quarrelsome, hence the wars that were endemic on the municipal level of politics. Ultimately, so the argument goes, they were to show a unique commonality of interests in the support they gave to the emerging absolute monarchies and nation-states of the late Middle Ages. Like the monarchy, they opposed feudal lords who placed imposts on their commercial transactions and blocked the development of their markets with a self-enclosed manorial economy.

The partial truth this conventional view conveys is outweighed by the serious error it contains. It expresses a characteristically liberal and Marxist prejudice that prevailed a century ago against all decentralized societies, a prejudice that was to be placed in the ideological service of European nationalism and its gospel of the centralized state. Only today do we seem willing to recognize how reactionary and false was this imagery of ever-embattled, quarrelsome, and promonarchical cities, whose economic power was presumably placed with few if any qualms in the service of absolutism.

There is more than enough historical evidence to show that cities were as disposed to form leagues and confederacies with each other as they were to fight with each other. Many of these leagues and confederacies, in fact, were not only networks of mutual aid; they were vigorously directed against absolutism and its threat to communal liberties. Finally, territorial lords were often quick to abandon their traditional feudal or manorial forms of rights and duties, to participate as vigorously in commerce as the most avaricious merchants. Indeed, English capitalism cut its first teeth in the countryside where many squires and nobles turned agricultural and common lands into sheep runs to meet Flemish demands for wool—perhaps the earliest example of agribusiness in modern times.

Confederacies or leagues of cities go back as far as Greek times when *poleis* entered into various associations with each other for mutual protection, shared religious beliefs, economic interests, even for quasitribal honorific reasons. At least fifteen of these confederacies of one kind or another, known more generally as *koinoi*, can be identified—many of which are very obscure but marked by fascinating examples of cooperation. These confederacies can often be traced back to tribal groups that were established as early as the Bronze Age. Tribalism never completely disappeared as a framework for the later confederations. Thus the famous Delian League that Athens developed was initially Ionian, composed of *poleis* that generally claimed a shared ethnic ancestry. By the same token, the Peloponnesian League that opposed it was largely Dorian, and the Achaean League claimed a shared ancestry with the archipelago's early Mycenean settlers although "Achaea" itself was really composed of a mixed population of Dorians and their precursors, the simpler Arcadians.

A troubling feature of many confederations is that one *polis* tended to become the pole around which its confederates clustered, whether by inclination, necessity, or coercion. The Delian League formed by Athens eventually became so overarchingly Athenian in character that historians were to call its later phase an "Athenian empire." This is an overstatement. That Athens battened itself on the revenues it extracted as "protection money" from the league and used coercion when persuasion failed to hold its confederates in line is doubtless true; but there is an inescapable irony in the fact that it foisted its own democratic institutions on *poleis* with limited freedoms of their own, whether they wanted a democracy or not. The internal politics of the league's members stands in very sharp contrast to the despotic institutions we encounter in virtually all ancient empires. In fact the original confederal council of the league, formed early in the fifth century to check Persia's advances into Greece, was distinguished by its high sense of fraternity. All members of the council had an equal vote, and its treasury was kept in the Temple of Apollo on the politically neutral island of Delos. Only later, when the Persian threat ended, did Athens assert complete sovereignty over the league, preventing Naxos and Thasos from seceding and bringing the league's treasury to the Athenian acropolis.

We also know of confederacies, however, where *poleis* were permitted to function very much on their own. Pellene, for example, showed considerable independence in the Achaean Confederacy; conversely, Thebes was held in check by *poleis* that made up the Boeotian Confederacy, particularly after it was reformed toward the end of the fourth century B.C. The use of the word *sympoliteia* to designate many confederacies, particularly those that extended beyond traditional tribal areas, is significant. Taken literally, the word describes a union of *poleis*, presumably of equal status, in contrast to a *patria* or "fatherland" with its connotations of a capital city, or an *ethnos*, with its real or fictive tribal bonds based on blood ties. At the molecular level of the *sympoliteia*'s life, the human bond is based on citizenship, on the *polites*, not on some form of juridical national identity at one extreme or kinship ties at the other. Citizenship, in effect, is not dissolved into an impersonal national affiliation or a presumably biological or tribal one. And, in fact, many *polites* or citizens of a confederacy enjoyed rights in other confederal *poleis* that they normally denied to resident aliens. They could buy land, enjoy the full protection of a confederated *polis*'s laws, and, in some cases perhaps, participate in its *ekklesia*, although normally their political rights were linked to the institutions of their own cities. In short, the confederacies of ancient Greece were to enlarge the whole concept of citizenship, well beyond the parochial framework of early *poleis*, while still maintaining their decentralized civic lifeways.

How were the confederacies structured? Our knowledge of them is very limited and any extended discussion of their known institutions is precluded by lack of space as well as facts. But certain general outlines can be noted. Normally, a Greek *polis* consisted of magistrates, a board of generals, a council or *boule*, and a citizens' assembly. This form of "government" existed in many Greek *poleis* from the fifth to the third century, when it was finally swept away by the Romans.

This is not an overstatement. The fact is that Greek *poleis* had very little experience with "representative" forms of governance; indeed it was very hard for the Greeks to think in terms of "representation" generally. They could understand the rise of an oligarchy, which they often identified with a tyranny or repressive control of some kind, and a democracy, which, in its Periklean form, seemed radical or "excessive" to its opponents. And republican-type structures did surface among them from time to time. But these republics rarely produced stable institutional forms. Human scale—a distinctly municipal scale—continued to be the only congenial and comprehensible level of institutional form that seemed to satisfy Hellenic lifeways and modes of thinking. Any form larger than the *polis* or confederacies of *poleis* cut across the grain of the Greek mind and Greek social theory. The rise of the Macedonian empire, regaled with all the trappings of royalty, generally horrified the Greeks, and the Roman empire seems more to have fascinated them, as Polybius's writings indicate, than attracted them until the very memory of political democracy had faded away. In any case, even when a republican regime did emerge among the Greeks, notably with the ascendancy of Rome, it was often called a *demokratia* and there was a tendency to trace its pedigree back to the *polis* of classical times. The Macedonian and Roman empires, in effect, constituted an annoying challenge to the Greek image of political consociation: its ethical as distinguished from administrative ways of visualizing or defining politics: its high regard for some degree of citizen participation in formulating policy or executing it. Just as a Roman in imperial times might look back nostalgically to the *republica*, so Greeks under foreign rule looked back endearingly to the *demokratia* and often used the word when it no longer applied to their institutions.

Not surprisingly, the *boule* and *ekklesia*—the council and the citizens' assembly—were to appear in many Greek confederations, not only within Greek *poleis* themselves. There is evidence that an *ek-klesia* formulated policies for the Thessalian Confederacy in the closing decades of the fifth century B.C., possibly as an aristocratic body. But in time cities began to encroach upon the power of the territorial nobles. Apparently, a democratic faction, strongly influenced by the Athenians, successfully extended popular rule within the Thessalian cities and the confederacy itself, after which the confederacy eventually became more centralized and exclusive. The Boeotian Confederacy was, as J. A. O. Larsen puts it, "a land of *hoplites*" because the area itself favored small-scale farming.[3] But it is surprising to find that the Confederacy more closely approximates a republican state than any we encounter so early in Greek history. This may have been the result more of Spartan influence than the internal development of the Thessalian *poleis*, an influence that did not go unchallenged by a pro-Athenian faction in the cities that made up the confederacy. Unfortunately, the details of its development are closed to us by the lack of adequate historical data.

The Phocian Confederacy alternated between an oligarchy and a democracy: a strong executive made up of generals had to answer to a popular assembly that enjoyed the power to depose its military leaders. The Locrian Confederacy seems to have had a citizens' assembly; indeed, from the scant evidence we have, democracy found a comfortable home here, often together with shared citizenship that made it possible for citizens of one *polis* to acquire property and to intermarry with citizens of another *polis*. In West Locris, the *poleis* were so indulgent that the Greeks generally singled them out for their fairly humane and decent treatment of foreigners. The Aetolian Confederacy appears to have had an *ekklesia* in which all citizens not only had the right to vote but followed the Athenian fashion of voting as individuals, not as citizens of their own *poleis*. This extraordinary degree of political individuation within a confederacy—a rarity even in decentralistic and confederal social theories—should not go unnoticed. What the Athenians did within Athens and its environs, the Locrians did within a confederation of separate cities. Meetings of their confederal assembly were held twice a year—once in the spring when the military campaigning season began and again in autumn when it came to an end. We shall have occasion to emphasize that democracy cannot be disassociated historically from military associations when they involve the mobilization of citizens for warfare, changes in arms and military technique, or simply a high valuation that is placed on the image of the armed citizen.

The Achaean Confederacy, perhaps the best known of all Greek confederacies, became so democratic that it was in advance of Athens in some respects. Finally, in 417 B.C., Sparta was obliged to step in to impose oligarchic rule. This intervention stirred up a medley of reactions in which a pro-Athenian faction restored democracy that led to further Spartan intervention. Nevertheless, there seems to have been a persistent *ekklesia* on a local level that certainly was in existence in Hellenistic times, the era following Alexander's conquests in the Near East and North Africa. Unfortunately, we know far too little about other Greek confederacies to provide even capsule descriptions about their structure and development.

What does it mean in very concrete terms to say that a Greek confederacy had a citizens' assembly? It is tempting to think that in comparatively large confederal areas, such an assembly is simply a euphemism for a representative system of government, not a direct, face-to-face body of citizens. Actually, this is far from true. Judging from the Achaean Confederacy, citizens from various *poleis* were expected to attend it *en masse*. For those days, this would mean a journey over wide distances, hence assembly meetings would tend to attract only the well-to-do who had the means and leisure to attend them. But much the same could be said of the Athenian *ekklesia*. Attika was more than the environs of Athens, and for communities in the more distant parts of Athenian territory a journey to the city would have been a fairly difficult one. Like Athens, however, the poorer elements in the host city or "capital" of a Greek confederacy often outnumbered the well-to-do who could afford to make the journey and may have provided it with a popular, indeed radical, ambience.

Whatever may have been the possibilities and limits of ancient cities, municipal democracy withered and finally died under Roman rule. The Roman Empire, a purely parasitic phenomenon, was extremely wary of municipal autonomy. It provided cities with only enough freedom to police themselves and extract tribute from subject populations. In the centuries following Periklean democracy, city life as a political reality began to decline and, after the second century of the contemporary era, shriveled disastrously, at least in Europe and the northern rim of the Mediterranean basin. Nor was urban life to revive in this area until the eleventh century. But with this revival came the emergence of new confederacies, an extremely important aspect of European history whose story has been badly neglected. Peter Kropotkin's work on the city confederacies and leagues of Europe, limited as it may be, may be cited as a truly pioneering effort.[4] The period of the French Revolution and the nineteenth century were to witness a depressing shift in perspective from historical studies of localism and urban confederalism to the nation-state, a shift that reflects a distinctly centralistic bias in radical as well as liberal historiography. The lacuna that exists in this field is by no means the result of oversight: it originates from a distinct political proclivity in Marxian historiography and liberal social theory to emphasize the role of the nation-state in fashioning the modern era, an emphasis for which we have paid dearly in evaluating the alternatives that face this era today with its increasing bureaucratization and centralization of social life.

Despite its brevity and incompleteness, Kropotkin's work still provides us with a robust framework for recovering some sense of the vitality this municipal world offered as an alternative to the nation-state. "Already in the years 1130–1150 powerful leagues came into existence," Kropotkin tells us, "and a few years later, when [Emperor] Frederick Barbarossa [of the Holy Roman Empire] invaded Italy and, supported by the nobles and some retardory cities, marched against Milan, popular enthusiasm was roused in many towns by popular preachers. Crema, Piacenza, Brescia, Tortona, etc., went to the rescue; the banners of the guilds of Verona, Padua, Vicenza, and Treviso floated side by side in the cities' camp against the banners of the Emperor and the nobles."[5]

The following year saw the emergence of the first of the Lombard Leagues (1167), which numbered sixteen cities at its height, followed by a second in 1198 and finally a third (1226) that collectively included nearly all the major cities of northern Italy. Not only was Milan a member of all three leagues but also Bologna, Verona, Brescia, Ferrara, Faenza, Vercelli, and Alessandria. Even Venice, proud and independent, joined the first of the leagues. A league of Tuscan cities was formed shortly after Henry VI's death in 1195, and still another two centuries later, guided largely by Rome during the papacy's quarrels with the empire. The number of leagues that formed in Italy during this time are too numerous to examine here. Some clustered around powerful cities such as Florence, Venice, Milan, and the papal seat in Rome, surfacing in the Romagna and in Umbria as well as in northern Italy. With the passing of time, these leagues either fell apart into rival cities or formed genuine city-states—in fact, small republics or duchies, depending upon the internal political structure of cities that led them. By the thirteenth century, this structure was usually oligarchical: the *popolo* had given way to *signori*, and Italy was to become a battleground for major European powers that tried to dip into the still very considerable wealth of the peninsula. Although eminent urban historians such as Lewis Mumford are highly disdainful of this development, particularly the continual discord between the cities that are believed ultimately to have fed the parcelization of the area, Kropotkin is careful to note that it was precisely "when separate cities became little States [that] wars broke out between them," generally as, a "struggle for supremacy or colonies."[6] Whether Italy's parcelization is quite the overall "evil" that characterizes most historical accounts of the Italian city-states or a desideratum that delayed the emergence of an overly centralized nation-state has yet to be assessed.

The creation of city confederacies in central Europe followed a development that is very similar to what we encounter in Italy, but they were also marked by characteristics that make them highly distinctive. That Italy led Europe in urban development is not surprising: the peninsula had been dotted by cities for centuries when much of the continent north of the Alps was still covered by forest. The German-speaking cities, however, were unique. Although they were to follow their Italian counterparts in time, they differed from them in social texture. They were burgher cities with sturdy domestic markets based on the ordinary staples of life to an extent we do not quite find in Italy (apart from Florence) or encounter in France. Cities involved in the Mediterranean trade made their fortunes largely from luxury goods such as silk, spices, gems, well-wrought armor, gold and silver ornaments, and the like, mainly transported from the Near East, North Africa, and Asia. By contrast, German cities tended to deal in the making and sale of coarser cloths, tools, simple armor, food staples, and raw materials. These commodities gave rise to a stay-at-home artisan and merchant order that underpinned very stolid communities with a deep sense of rootedness and a strong appetite for security. Accordingly, a localist civicism and proclivity for autonomy persisted after the Italians had become relatively jaded in their municipal loyalties and yielded to despotic regimes. The German word *Gemeinde* has a special meaning in civic history for which other languages have poor equivalents. It denotes an organic community, a community that has a sense of identity and personality, indeed one in which city hierarchies are notable for the contribution they make to the collective good at each level rather than the oppression they inflict on subordinates.

Genoa and Venice acquired their wealth mainly from exotic goods and a Mediterranean-wide trade. Hamburg acquired its wealth from brewing and Lübeck from herring and the furs of east European forests. Merchant and artisan, trader and primary producer developed a symbiotic relationship that was relatively rare in the Latin cities of the south. The city confederacies projected their burgher traits onto their confederacies: cities and towns came together not only to protect their autonomy and liberties; they also joined to promote trade and share in a common prosperity, not simply as rivals whom circumstances forced into collusion with each other. The persistent conflict that marked so many Italian cities, especially when they developed into city-states that placed lesser communities under their control, was more subdued north of the Alps. Although bitter internal wars unravel this picture in the Flemish cities, where a nascent "proletariat" stood at loggerheads with a nascent "capitalist class" in the wool-processing industry of the time (a problem, I may add, that afflicted Florence no less than Bruges), the guild structures of central and northern Europe were more entrenched than elsewhere. They helped to create and empower a stratum of middling people, mainly artisans and small merchants, who enjoyed relatively comfortable lives and had a stabilizing effect on the community, cushioning the conflicts that were spawned by great disparities of wealth.

Thus, one has the sense that German cities formed more stable confederacies than did other urban entities in Europe. Indeed the Swiss Confederation, perhaps the most enduring and libertarian to emerge in Europe, rested heavily on the formation of the *Graubünden* or "Gray League," the canton that was to be dubbed *die kleine Schweiz* or "little Switzerland," partly because of its prototypic character as the home of Swiss democracy, partly too because of its ethnic diversity although its population is mainly German-speaking. Here, the Swiss recourse to referenda is reputed to have been born and "Nowhere through the whole range of history," declared F. B. Baker exultantly, nearly a century ago, "is it possible to find a country where the democratic principle was more thoroughly applied ... or where the good and bad results of that principle have been more thoroughly demonstrated."[7]

Mumford's churlish statements about the municipal confederacies of central Europe to the contrary notwithstanding, what the cities of Germanic Europe lacked in durability, they tended to make up in recuperability.[8] Some four centuries of German history are marked by a large number of municipal confederations that continually bubbled up to the surface of political life. The Hanseatic League, perhaps the most durable of the lot, existed from 1241, when Lübeck and Hamburg signed a treaty of mutual protection, to 1669, when its last diet was convened. Officially, the league was never terminated and cities such as Hamburg and Bremen are still designated as "Hanseatic cities." Largely based on the Baltic trade, the league at its height embraced between 60 and 80 cities (I have taken the most conservative figures at my disposal), including the wool-processing center of Bruges in Flanders. Nearly all the major Baltic ports belonged to this confederacy at one time or another, and its ships ranged widely from

Novogorod in the east to London in the west and along the Atlantic coastline.

Still earlier, major confederacies appeared in parts of central Germany, principally the short-lived First Rhenish League in 1226, followed by the Second in 1254, which lasted until 1258. Some eighty cities, virtually all the leading Rhineland communities, belonged to the league until its members drifted away after supporting contending claimants to the throne of the Holy Roman Empire. Intermixed with the politics of the empire and endemically at war with nobles who preyed on their trade, their history weaves a story of enormous complexity and challenges. By 1384, a Swabian League had been formed that brought German cities to unprecedented influence. League members even advanced proposals to join the Swiss Confederation. Had the Swiss been responsive to these overtures, European history might have taken a very different turn than it did, possibly replacing nationalism with confederalism. But the union was not to be, and the cities, ever mindful of their autonomy and liberties, failed to prevail over the empire and the princes. Later leagues were to appear throughout Europe, even in England and, very significantly, in Spain. If we bear in mind the large number of municipal confederacies that existed in Europe during the eleventh century and in the centuries that followed it, the certainty so prevalent in present-day historiography that the nation-state constitutes a "logical" development of Europe out of feudalism can only be regarded as a bias, indeed a misuse of hindsight that verges on a mystical form of historical predetermination.

Again, how, may we ask, were the Italian and northern European municipal confederacies structured? None of them created the popular intercity assemblies we encounter among the Greeks. Although citizen assemblies emerged within the cities, they did not appear between the cities. In Italy, the ad hoc nature of the confederacies did not create any serious problem of entrenched representatives who could defy the will of their constituencies: the confederacies were notable for their impermanence. They were little more than defensive alliances and disappeared as soon as they were not needed. The First Lombard League created a "parliament" of its own, but as

Daniel Waley tells us, it simply "assigned to its members areas of military responsibility and settled the price contributions of each to field armies (*tallia milium*) and garrisons." This "parliament" never became a supracommunal authority." Rather it functioned more like a temporary general staff. "To the communes," Waley adds, "this *societas* or League was, like the Empire, an institution to be judged by its utility rather than by any theoretical implications: only an immediate imperialist threat could keep it in being."⁹ To the extent that we can speak of "capitalism" at this time, the highly aggressive entrepreneurial spirit of the Italian municipalities, fed by the enormous wealth of the Mediterranean-wide trade, fostered a degree of rivalry that inhibited cooperation between the cities and imparted a highly imperialistic spirit to the largest among them.

This was not entirely the case north of the Alps. German city confederations, for example, provide us with more enduring efforts to institutionalize intercommunal cooperation, efforts that reflect the prudent and deeply rooted burgher spirit of the communities that composed them as distinguished from the more reckless and venturesome features of the Italian merchants. The Second Rhenish League, shrewdly playing off the different candidates for the imperial throne, demanded and received formal recognition from William of Orange as a confederation—a *civitates conjuratae*—and avowed in its declaration that its citizens "have mutually bound [themselves] by oath to observe a general peace from St. Margaret's Day (July 12, 1254)." This declaration was to go well beyond peacekeeping. Each member of the league agreed to send four representatives to a city assembly or *Städtetag*—one at Worms for the upper Rhine, the other at Mainz for the lower—to remove excessive river tolls, provide for the common defense, add or expel new members, and foster the commerce and welfare of each of the league's members to the benefit of all civic orders, including Jews and clergy, not only ordinary citizens. A board of arbitration was established to settle quarrels between confederate cities. Finally, an assembly meeting of the municipalities' representatives was held on a quarterly basis, not on the usual annual one we encounter elsewhere.

The Swabian League followed almost naturally out of its Rhenish predecessor and functioned as a countervailing force to the empire and the territorial lords. Its very formation without imperial sanction was an act of defiance against the efforts of Emperor Charles IV to assert centralized control over the German cities, hence its articles of agreement have a markedly defensive tone. But it functioned very much like the Rhenish League. By the 1380s, the confederacy forced the princes into temporary submission and, in combination with the League of the Rhine, which it structurally resembled, formed one of Europe's greatest urban confederacies.

The emergence of the celebrated Swiss Confederation or "Switzering anarchists" as Cromwell's supporters were to call it centuries later, must be seen as an extension of the Rhenish and Swabian leagues, not an anomaly that stands at odds with the supposedly parochial traits imputed to European cities and their leagues. Switzerland was formed out of a milieu and modeled after examples that existed in central Europe as a whole.[10] The Swiss confederation, far from being an almost lonely "exception" to the confederal trends that existed in the thirteenth and fourteenth centuries, was actually a product of them. That the creation of nation-states was to be so greatly delayed in Germany and Italy is due in great measure to the obstacles that the cities, their confederacies, and later the powerful impact of their traditions of autonomy and freedom exercised on political life as a whole. It was not localist "pettiness" and "parochialism" that kept central and southern Europe from achieving nationhood until well into the nineteenth century. Rather, the "delay" was in great measure the product of a strong tradition of municipal autonomy and a dramatic history of resistance to centralization, however perverted this history became in later times. The Hanseatic League's Diet, the *Städtetag* of the Rhenish League, and autonomous confederal bodies elsewhere in Europe haunt the history of the continent like the unexorcised spirit of a more active public life and a vital civic politics. That the nation-state eventually did unite the laggard principalities of Germany, Italy, and, as we shall see, Spain into centralized states was not quite the happy dispensation it seemed to be at the turn of the present century, when nationhood was regarded as evidence of "modernity" and "progress." Viewed with hindsight, the images of Mussolini, Hitler, and Franco rise up to remind us that the ideological celebration of the nation-state, which marked social theory during the Victorian era, was grossly misplaced. We of a later generation have good reason to lament the loss of the confederal alternative that appeared at an earlier epoch in Europe, one that might have averted the terrifying turn "national unity" took between 1914 and 1945.

Confederalism was not merely an intuitive civic reaction to the feudal parochialism that marked so much of the medieval world at the time. There were theoretical not only practical considerations that were to surface almost simultaneously in two charismatic figures of western Europe. Heraldic rebels such as Cola di Rienzi in Rome and Etienne Marcel in Paris, contemporaries during the mid-fourteenth century, were to formulate the numerous moves toward confederal unity in very dramatic form. Rienzi's efforts to restore a new Roman republic evoke images of the Gracchi and their efforts to restore Rome's traditional republican virtues. His attacks on the venal nobility of papal Rome and his efforts to create a citizens' militia were apparently part of a larger effort to unite Italy into civic leagues under Rome's suzereinty. Present-day historians tend to depict Rienzi as a forerunner of Italian nationalism when they are not occupied with defaming him as a "demagogue" with strictly self-serving intentions. The greater likelihood is that he was a strident leader of Italian confederalism. As a self-styled "Tribune of the People," a title redolent of the Gracchi rather than a Cincinnatus, Rienzi's "parliament" was to be made up of delegates from Italian cities, not peninsular "provinces" that had yet to come into being. This effort was to be aborted when the papacy and nobility allied with each other, ultimately leading to his murder in 1354.

Etienne Marcel emerges from this stormy era in a far more favorable light. A "provost of the merchants" and economically well-to-do in his own right, Marcel was clearly a popular leader of the "Third Estate" in Paris whose efforts to enlarge the powers of the *Estates Generale* at the expense of the monarchy and nobility and make royal taxes more equitable developed into a wider challenge to absolutism and aristocratic power. Marcel's own goals, in fact, were probably more "bourgeois," possibly even more "republican," than his own order would have been prepared to accept. Like Rienzi, he was to enjoy immense support until a royalist faction within the middle class itself assassinated him in 1358. The tendency of nineteenth- and twentieth-century historians to read a "nation-state" mentality into men who lived in an era guided largely by feudal or confederal visions of political life is misleading. Underlying this bias is the myth that Europe's "bourgeoisie" was initially republican and basically nationalist in its convictions. Rierizi and Marcel rose to prominence because they spoke for artisans and the urban poor as well as merchant and professional strata, many of whom sided with the nobility against the ordinary people of Europe's cities. Perez Zagorin seems to be much closer to the truth when he observes, "Grievances born of unfavorable *conjoncture* and wretched conditions underlay the popular upsurge, whereas Marcel himself was a revolutionary reformer who wanted to build an alliance of towns, strengthen the Estates General, and fasten political limitations on the monarchy. His movement also established some slight ties with the Jacquerie, the big peasant revolt that had broken out at the same time in the Ile-de-France and surrounding region."[11] This judgement speaks to a confederal outlook, one that was more in tune with the period than a "nationalist" vision of a centralized France.

Zagorin's reference to the Jacquerie reminds us that the period faced a series of major peasant tumults—not only in France but throughout western Europe. The English Peasant Rebellion of 1381, which followed the French *Jacquerie* by less than three decades, formed a high point in the restive village upsurges that finally led to the brief seizure of London by a peasant army. John Ball, an itinerant priest who was to color the rebellion with powerful declamations against social and economic inequality, gave the rising a larger-than-life image. Actually, the peasantry tended to rise in the aftermath of wartime ravages of the countryside or unpredictably in isolated pockets against feudal exactions, royal taxes, excessive "tithes," upper-class highhandedness. To detail these uprisings would be impossible. Occasionally, they managed to cojoin with urban unrest into full-scale rebellions and to pose serious challenges to provincial and central authorities. Whether this connection was well-established—as was the case in the Hussite Wars of Bohemia during the fifteenth century—or assumed an ephemeral character, agrarian unrest became endemic throughout the centuries that followed and persists to this day in the Third World.

Ordinarily, peasant revolts were short-lived and fragmented. The climax of these uprisings in western Europe was to emerge from the Lutheran Reformation of the 1520s, when the cities and countryside of Germany were thrown into years of persistent unrest. Between 1524 and 1526, German peasants rose on a mass scale; sweeping over large areas of the western and southern parts of the country and entering into historical annals as the famous "Peasants' War." Presumably, this was regarded as a "revolution" by chroniclers, who were eager to distinguish it from the many "rebellions" that exploded sporadically in the countryside from medieval times to the Enlightenment. The medley of ideologies, sentiments, and interests that are imputed to the German peasant uprising has endeared it to Marxists, liberals, romantics, theologians, and nationalists alike. It has been variously seen as a precursor of modern communistic movements, a striking example of class war, an effort at moral regeneration, even as a forerunner of the German nation-state or a force that shaped its development. Most of these appraisals do not give sufficient emphasis to the deep-seated communitarian impulses that moved the peasants to action: their attempts to preserve the rural *Gemeinde* from feudal, commercial, and clerical encroachment. If there is any unifying drama to the upheaval, it is the peasantry's effort to preserve its organic communal ties, its traditional village universe that encompassed time-honored values, institutions, and lifeways as well as landholdings that were challenged by princely and baronial encroachments. It is this universe, so much of a piece with the civic values of traditional society as such, that makes the Peasants' War of 1524–26 so fascinating to theorists and historians of municipal development.

The fortunes of this conflict, with all their varied interpretations, have been explored extensively enough to require no detailed treatment here. Thomas Munzer's legendary "communist" tenets of the time probably articulated the peasantry's commitment to its traditional networks of mutual aid, its timeless visions of a "golden age" based on equality, its precarious reality of collective management of land and goods that marked the *Gemeinde*—all village-based traditions rather than "anticipations" of socialist and communist theories that appropriately stem from a sophisticated industrial society rather than an old agrarian one. To the romantics who found in the peasantry the embodiment of a German *ethnos*, the conflict offers no inspiring myths of a united people motivated by a sense of blood and soil. The Peasants' War was as fragmented as the society from which it stemmed. Only once do we hear of a really earnest effort to bring what were separate uprisings into a unified struggle. A "peasant parliament" was convened at Memmingen to form a "Christian Union of the Peasants" and coordinate the peasant armies of Upper Swabia. The union brought together the three *Haufen* or corps that were in revolt into common military operations, each led by a chief and four councilors, a structure that completely replicated the administration of the peasant village. The village form, in effect, was projected onto the shared command structure of the military forces, mirroring the traditional society that it was meant to preserve.

The Memmingen "parliament" also formulated and adopted the famous "Twelve Articles" of the peasant revolt, a program worded in terms of scriptural authority. The articles "humbly" petition the secular authorities for the right of the villages to choose and depose their own pastors, to fix their own "tithes" according to the needs of the pastor and the poor of the community, to abolish serfdom, diminish corvée labor, reduce feudal dues and rents, and, finally, to restore all enclosed common land to the village and end further enclosures definitively. Allowing for many local variations, the Memmingen Articles became the basic program of the rebellious German peasantry and was soon the most widely circulated document of the war. Given its tone of humility, recourse to scriptural legitimacy, and humane demands, it completely expressed the spirit and values of the *Gemeinde*. It was the voice of a traditional village world by which the municipal life of the era was nourished and from which it drew its vitality. Herein lies the real continuity of the articles with German civic life: its strong solidarity as an ethical covenant.

The German princes were to unite and crush the Peasants' War in a terrible bloodbath. Although city support of the peasants was very widespread, it was often qualified and prone toward compromise. The Twelve Articles resonated with the urban poor and lower classes, and its moral tone and appeals to scripture won it considerable clerical support. The almost evangelical nature of the uprising gave it the qualities of a crusade for human rights and decency, traits that were not lost on many of the educated strata of the cities. Despite its radical rhetoric, Friedrich Engels's appraisal of the Peasants' War is pervaded by all the prejudices of the last century. The war's "chief result," we are told, was the "strengthening of German decentralization" which, in turn, was the "cause of [the war's] failure."[12] One is disposed to ask if victory by the peasants would have yielded a "centralization" of Germany that would have overcome the fragmentation created by the German princes. Accordingly, the Peasants' War was either a "revolution," as Engels claimed, or a crude anachronism as he should have claimed by his own standards of "historical materialism," in which case it belonged not to the truly "revolutionary tradition" of Germany but to a "reactionary" one. His appraisal becomes all the more entangled when he rejects confederacy as a valid solution to Germany's problems, a solution that the peasants intuitively seem to have desired. This problem is not an academic one. It raises the crucial question of whether or not seemingly "undeveloped" peoples today are to achieve what we so flippantly call "modernization"—by confederalism or nationalism, decentralism or centralism, libertarian institutions or authoritarian ones. We have not removed these questions from the future of our civilization nor can they be concealed from purview by the veil of history. If anything, hindsight has made them as searing today as they were in earlier times, when the terrifying future that now looms before us was very far removed from the eyes of men and women in the sixteenth century.

This much must be emphasized: the attempts to create a nation-state in western Europe four centuries ago did not occur without considerable resistance from the free cities of the era, rebellious villages, and aroused artisans, not only recalcitrant nobles. The sixteenth century, which was decisive in the rise of European absolutism and ultimately nationalism, bears witness to a veritable deluge of village, provincial, and urban uprisings. Even England was not spared from dramatic agrarian unrest: Kett's Rebellion of 1549, while more of a mass protest against land enclosures than a revolutionary challenge to royal authority, required the use of thousands of troops before it could be subdued. In France, the rebellion was followed by the rising of the *Croquants* between 1592 and 1595, a struggle more redolent of German demands for village autonomy than the English one. Such revolts, generally localized and easily subdued, were to become endemic as France passed deeper into absolutism and ultimately revolution. Indeed, it was not until the Napoleonic Era that they came to a definitive end, and a once rebellious peasantry was turned into a conservative pillar of the Bonapartist monarchy.

None of these rebellions produced confederations or developed into serious challenges to the emerging nation-states of the West. The civic roots of the English Revolution have rarely been appraised from a municipal viewpoint, although the revolution that began in England in the 1640s was to find a remarkably democratic fulfillment in the townships of New England. (This is a development we must reserve for a later and fuller discussion.) The Great French Revolution, in turn, was to evoke the ideal of communal confederation without giving it permanent reality. Indeed, the Jacobin "dictatorship," if such it can be called, was to turn France into one of the most centralized nation-states in Europe. But the ideal did not die. Later, it acquired a brief and glowing moment of reality in the subsequent Paris Commune of 1871, a commune or "city council" that called upon all the cities of France to join it in a huge civic confederation—only to be crushed in bloody fighting with troops of the Third Republic after some two months of existence. With magnificent stubbornness, the Paris Commune of the nineteenth century had tried to bring to life what its predecessor of the eighteenth century had entertained in its conflict with the Jacobin-controlled Convention during the last years of the Great Revolution.

Ironically, the most serious threat to absolutism and the nation-state by a confederation of municipalities was to emerge during the sixteenth century in a country where absolutism seemed virtually triumphant: Spain under the rule of Europe's sternest and politically least yielding monarch, Charles V. The period directly preceding the final subjugation of the Moors in 1492 was marked by a remarkable burgeoning of city life and the consolidation of Spanish absolutism under the Catholic monarchs, Ferdinand and Isabella. Spain's prosperity and its movement toward a nation-state seem like an almost textbook example of collaboration between a monarchy and an urban "bourgeoisie." That monarchy and city could have eventually entered into conflict with each other would have seemed inconceivable at the time if it did not actually happen. Isabella, following an active policy of fostering city life and playing urban strata against landed magnates, promoted internal commerce, scrupulously respected municipal rights, and worked closely with the Spanish *Cortes*, the country's "parliament." This policy was intended to develop institutions and forces countervailing the rapacious Spanish territorial lords. The guilds, long the objects of royal hostility, were permitted to extend their ordinances and confirmed in their rights to control production operations. The "crusade" against the Moors was shrewdly used to increase popular enthusiasm for the monarchy, indeed to give it a centrality in Spanish life and religion that it had never enjoyed. The image of a morally regenerated Christian Spain evoked hopes of a stable, unified, and powerful nation whose restive nobility and centrifugal regionalism had to be brought under rigorous control.

This image was translated into a certain measure of fact, especially in Castile, the Spanish heartland. Castile formed the bulwark of the newly emerging nation, the source of its prevailing dialect, its manners, and the "prototypic" Spanish character, which held such linguistically, ethnically, and culturally disparate "Spains" as the Basques, Catalans, and largely Moorish Andalusians together. Here, too, the monarchy in later years was to choose its future capital, Madrid, and turn the city into the administrative center of the country as a whole. This state machinery was one of the most sophisticated in Europe. At the same time, the Catholic kings began to shore up their relationship with the cities, deflecting urban hostility from the monarchy to the nobility and drawing upon urban wealth to consolidate royal rule. Ferdinand and Isabella enlarged their bureaucratic control over the independent cities of the province. The *corregidores* or town officials of the crown were given extended powers to bring the urban noble clans under control and protect the cities from landed magnates who exploited them. Municipal land that the magnates had illegally seized were restored in some measure; tax farmers, whose actions verged on outright plunder, were replaced by the *encabezamiento* system in which the main tax, the *alcabala*, was collected by local officials; the Royal Council was staffed by university trained lawyers, the *letrados*, and a supervisory hierarchy of officials to oversee the burgeoning bureaucracy from secretaries to *visitadores* who provided for redress from abuses and grievances. A centralized and professional judiciary, together with various councils of brotherhoods, the Inquisition, and the *Cortes* itself balanced out a bureaucracy and provided for close royal supervision of nearly all aspects of Spanish life.

This machinery, partly traditional and partly new, was destined to have a very limited life span. Although the French monarchy was to install a similar one that lasted for some two centuries, the Spanish state machinery began to weaken appreciably even during the reign of the Catholic monarchs. The final struggle against the Moors essentially brought the conflict between the monarchy and nobility to an end. Despite her fears of the landed magnates, Isabella was obliged to use them militarily to complete the reconquest, and the magnates claimed their full rewards for supporting the state. Increasingly, the lost municipal lands were recovered by the aristocracy, their taxing powers were increased, their abilities to sidestep disagreeable court decisions were enhanced, and their financial control of the monarchy increased immensely. When in 1519 Charles V became king of Spain, as Carlos I, and entered his claim to the throne of the Holy Roman Empire, the monarchy was largely under aristocratic control. Born and raised in Flanders, Charles quickly earned the mistrust of his Spanish subjects as a foreigner whose principal concerns were his own imperial ambitions and who lived mainly abroad to advance them. Virtually all strata of Castilian society viewed the newly installed king as a man who regarded Spain as a resource for his squabbles abroad, freely bilking the country of its wealth.

To exacerbate matters still further, the central administrative apparatus was all but taken over by a coterie of Flemish advisors that was notable for its insensitivity to Spanish traditions and interests. Inept and clumsy, Charles's Flemish surrogates turned the monarchy into a parasitic entity in which the aristocracy was the main beneficiary of the new dispensation. Increasing taxes, a grave decline in the honesty and effectiveness of the *corregidores*, a breakdown in the road system that led to higher, often unbearable financial levies on impoverished village and town populations, a failure by inept or corrupt supervisory officials to discharge their responsibilities in controlling the bureaucracy and aristocracy, a decline in the integrity of the court system, a military system that quartered ill-paid and unruly troops with an increasingly impoverished population—all, well underway before Charles became king—were exacerbated by alien rulers, a suspect monarch, the declining prestige of the royal power, and the growing encroachments of a self-serving aristocracy.

On May 30, 1520, a crowd of woolworkers seized a hated member of Segovia's *Cortes* delegation and hanged him, leading to a revolt in the city that forced all its royal officials to take to their heels. This seemingly local act of crowd violence was to unleash one of history's most extraordinary municipal "revolutions," as many historians call it, the rising of the *Comuneros* (literally translated as Communards). Although this *comunidad* or community revolt was fairly short-lived, it is outstanding for its institutional creativity. The action of the woolworkers in Segovia pales before the more serious rebellion that developed when Toledo's city council, challenging an unfavorable change in royal tax policy, wrote to all the cities represented in the *Cortes* and defiantly called upon them to establish a common front against the royal government. What may have appeared like one of many urban tax revolts that marked the whole period soon turned into a full-scale revolution. Within months, city after city in Castile began to collect and impound all taxes collected for the monarchy. Civic militias were organized, and far-reaching changes were introduced to democratize and enhance the autonomy of municipal governments. On Toledo's suggestion, a national *junta* was established with delegates from all the *Cortes* cities. The *Comuneros*, in effect, had established a parallel or "dual" power in opposition to the prevailing royal administration.

Early reactions to this development ranged from the enthusiastic to the tolerant. Even the landed magnates, ever mindful of an opportunity to gain from any diminution of the central government, placed a tactful distance between themselves and the monarchy. By mustering an impressive army of citizens with an infrastructure and added detachments of professional soldiers, the *Comunero Junta* moved speedily toward a series of victories that threatened to replace the entire royal state with a municipal confederation. The *Comuneros* had created their own military system, an administrative apparatus that reached deeply into most of Castile's social order, tax resources, and a tremendous reservoir of popular goodwill, cutting across seemingly insurmountable "class" barriers (including clerical ones) that seemed irresistible in the early months of the *Junta*'s existence.

What brought this movement to an end in April 1521 when its last major field detachments were defeated near the village of Villalar? Toledo, it should be noted, held out against royalist forces until February 1522, and other cities tried to resist after the battle of Villalar. Perhaps the most strategic military fact was the swing of the nobility from a generally neutral position over to the monarchy. No less important was the support that the royalists slowly acquired from the city elites—the knights or *caballeros* who lived in urban areas, well-to-do merchants, the higher clergy, and generally more prosperous strata, who were alienated by the radicalization and democratization of civic life. The *Comuneros*, like their heirs centuries later in Paris, were stridently urban in outlook and retained a basic hostility to the peasantry (who actually were their natural allies) as a class controlled by the nobles. Finally, the *Comuneros* could not extend their movement beyond the center of Spain. Viewed as Castilians by the other "Spains" that surrounded them, the movement was seen as the work of a privileged population that had revolted against its even more privileged overlords. The Catalans, Basques, and Andalusians, to cite the most well-known regions hostile to Castilian hegemony, could not be brought to identify with a Castilian cause, however much the *Comuneros* solicited their support

It is easy in view of these reasons to see the revolt of the *Comunidades* as a strictly "class" movement, to speak of it variously as "atavistic" because it posed a mere municipal challenge to a seemingly "progressive" nation-state or to regard it as a conflict of interests between vague, indefinable class strata: nobles, a "bourgeoisie," a "nascent proletariat," and the like. The term "nascent" is what makes such a "class analysis" questionable. Of all the clearly definable "classes" that were to play a major role in later Spanish history—apart from Spain's enduring peasantry—only the landed magnates survived the era as a cohesive stratum and were to carry on intact until recent times. The others are more properly "orders" in their indefinability, that is, in their paucity of economic roots, their wavering stability as social groups, and the murkiness of their concerns. What we have here is a typical, quasifeudal "Third Estate," ranging from well-to-do, even wealthy, strata to an amorphous mass of artisans, merchants, "intellectuals" of various sorts, clerks, and clerics, to which we may add a considerable number of servants, laborers, and beggars. This "Third Estate" was united by its urbanity, literally by a shared culture as town dwellers. Despite the many material differences that were to separate them, either they were citizens of a particular city or they aspired to be. Their ideological unity came from the primary loyalty that the city claimed and from the political arena it created. It was the city, not their "class," that evoked in them a real feeling of place, a meaningful commitment of service, and a clear sense of self-definition. This collective loyalty to a *patria chica*, to a "small fatherland," so intense among urban dwellers during that era, is difficult to convey today when nationalism has invaded all public sentiments of local loyalty. In the sixteenth century it was intense enough to impart an alien, almost exogenous, quality to the central power and to focus one's devotion on the village, town, or city in which one lived rather than the still-emerging nation-state.

Nowadays, we would be inclined to believe that such varied economic groups would be in chronic conflict with each other, a conclusion that seems to be supported by the internal conflicts that engulfed many cities of Europe, particularly the Flemish and Italian ones. Actually, this is a very one-sided picture of urban life in the past. It is easy for historians to forget how readily disparate strata in a city united against invaders or other city rivals, despite their divergent economic interests. In fact, it would be hard to understand why the *Comuneros* could unite in the first place, given the disparities in wealth and social position that existed in their cities, and why a strong sense of unity existed to the very last, even after urban elites began to fall away from their movement. Their royalist opponents did not win all the well-to-do strata of the Castilian cities; in fact, there was resistance to the very end, especially in Toledo, which held out against royalist opponents for nearly a year after the battle of Villalar. What the royalists succeeded in achieving was enough of a division between these strata to tip the balance in their favor and bring the greater military prowess of the aristocracy into an advantageous position over relatively inexperienced civic militias.

What the *Comuneros* really achieved has yet to be fully grasped by some of its historians. The movement opened civic life on a scale that had rarely been seen in Europe since Hellenic times. It expanded the very meaning of the word "politics," not only at a confederal, city, or town level but at a neighborhood or parish level. *Comunero* demands were strikingly radical even for our day: a *Cortes*, composed of city delegations, which would greatly limit royal authority, and a municipal democracy whose extent varied from one city to another. In a group of articles formulated in Valladolid, the *Comuneros* demanded that delegates for the Cortes be chosen with the consent of parish assemblies instead of city councils, the practice followed by the monarchy. These delegates in turn were to be guided by the mandate of their electors and were to acquire the right to consult with their cities if their instructions did not adequately cover the problems that surfaced at the Cortes, a right that the monarchy had consistently denied city delegates to the parliamentary body. Had these demands been realized, Spain would have seen the emergence of a broadly based local democracy, one deeply rooted in city neighborhoods as well as municipalities. Such a democracy, in fact, went far beyond radical conceptions of political representation. They were an open invitation to revitalize the entire public sphere, opening it to all strata of the population and advanced urban concepts of citizenship that were all-inclusive and completely grassroots in character. In cities such as Toledo and Valladolid, this neighborhood democracy was not merely a demand; it became a working reality, one that was rarely to be achieved again until the rise of the Parisian sectional movement in the Great French Revolution.

Many *Comunero* demands constituted a sixteenth-century "Bill of Rights." The *Cortes* was to meet regularly and all the grievances of the *Comunidades* were to be addressed before it could terminate its proceedings. The *Comuneros*, of course, called for the protection of property from legal confiscation except in cases of treason; freedom from harsh punishment in criminal cases; limits on the quartering of visiting royalty; prohibition of the sale of public offices; reforms of judicial and appeals procedures; and the complete "Castilianization" of the court, which Charles had filled with aliens who knew very little about Spanish problems. The demands contained a strong, basically egalitarian appeal for choosing officials according to their personal merit, professional qualifications, and moral probity rather than for their status and social background.

Charles's victory over the *Comuneros* signaled the triumph in Spain of statecraft over politics, of the nation-state over confederalism. It was a victory that was attained primarily by the force of arms, not by a hidden logic of history. The struggle of the *Comunidades* with the monarchy—it was never a struggle against monarchism as such although it came very close at times to a challenge of monarchical rule in its sixteenth-century form—had been preceded by similar conflicts between city leagues or confederacies almost everywhere on the European continent. It was to be followed by greater or lesser struggles of a similar nature after nation-states had been well-established. If Spain, one of Europe's strongest absolute monarchies in the sixteenth century, is singled out for study, the *Comunero* movement did not establish a tradition that an "ascending" bourgeoisie could claim for itself. Quite to the contrary: the *Comuneros* found a later, albeit highly modified, expression in Pi y Margall's Federalist movement of the late nineteenth century, which distinctly resisted state centralization, and finally in the largest anarchist movement in Europe.

Charles V did nothing to foster a capitalist society. Indeed, absolutism became a lethal cancer in a once prosperous country that was devitalized by massive state expenditures for imperial adventures abroad. The *Comunero* movement, by contrast, tried to rein this monarchy and ultimately drastically diminish its power. Its failure was followed over a period of time by an incredible decline of Spanish economic and urban life. Cities sank into lethargy, agriculture stagnated under the rule and mismanagement of the magnates, roads were permitted to decay, and the wealth of the country was vastly diminished. On the other hand, the Industrial Revolution in Europe, which presumably dates the ascendancy of urban capitalism over traditional society, did not foster a city development in Spain that was wholesome or vital. It did not revive community life; rather, it replaced what remained of community with urbanization, anomie, and, under Francisco Franco, with a ferociously terrorist regime that has variously been called "nationalist" by its admirers and "fascist" by its opponents. Whether or not the two terms actually reveal the convergence of a development that was to yield centralized authority in its most brutal forms is a problem that we have yet to resolve in our own time.

That there is a logic in certain historical premises, one that unfolds more as a tendency than a necessity, is certainly not arguable: nationalism does foster totalitarianism, and the centralized state tends to develop into an all-embracing state. But it is certainly difficult to argue that a superhuman phenomenon called "history" exists and predetermines a society's development. The *Comuneros* had opened a pathway to a cooperative, unified Spain that could have yielded a very different dispensation from that which came with a centralized nation-state. So, too, had earlier city confederacies, whose achievements meet with so much disdain. Politics had to be structured around a community of one kind or another, whether as a *polis, Gemeinde, burg, commune,* or city. Lacking the flesh and blood of politically involved people and comprehensible self-governing institutions, the human phenomenon we call "society" tends to disintegrate at its base, even as it seems to consolidate at its apex.

Centralization becomes most acute when deterioration occurs at the base of society. Divested of its culture as a political realm, society becomes an ensemble of bureaucratic agencies that bind monadic individuals and family units into a strictly administrative structure or a form of "possessive," more properly acquisitive, individualism that leads to privatization of the self and its disintegration into mere egoism. The city, in turn, is no longer united by any sort of ethical bond. It becomes a marketplace, a destructured and formless economic unit, a realm in which the Hobbesian war of "all against all" becomes a virtual reality, ironically designated as a "return to nature."

Such a condition and the mentality it produces constitute a dissolution of nature and society's evolutionary thrust toward diversity, complexity, and community, a problem that appropriately belongs to the newly developing field and philosophy of social ecology rather than urban sociology. It is a social problem because we are talking about one of the most elemental forms of human consociation—the city—where people advance beyond the kinship bond to share, create, and develop the means of life, culturally as well as economically, as human beings. Here, *humanitas* as distinguished from the "folk" comes into its own. And it is an ecological problem in the sense that diversity, variety, and participation constitute not only the basis for the stability of human consociation but also for the creativity that is imparted to us by diversity, indeed, ultimately, the freedom that alternative forms of development allow for the evolution of new, richer, and well-rounded social forms.

Urbanization, which I see as the dissolution of the city's wealth of variety and as a force that makes for municipal homogeneity and formlessness, is a threat to the stability, fecundity, and freedom that the city added to the social landscape. A critical analysis of how urbanization emerged, its genesis partly in the nation-state, partly in industrialism, and generally with the onset of capitalistic forms of production and distribution—all examined from the viewpoint of social ecology—is a problem of crucial importance for our time, indeed one that will help us define the future of the city, politics, and, above all, citizenship.

Chapter Seven
The Social Ecology of Urbanization

From the sixteenth century onward, Europe was the stage for a drama unique in history: the development of nation-states and national cultures in which populations tended to identify with what we, today, accept as a commonplace—a sense of personal nationality. Even the notion of citizenship, long-rooted in loyalty to a city and the public body that occupied it, began to shift toward a large territorial entity, the "nation," and to its "capital" city. Politics, too, began to acquire a new definition. It increasingly denoted the professionalization of power with roots in the state and its institutions.

We would be gravely mistaken to assume that these changes and redefinitions occurred within the span of a few centuries or that they have been completed even in our own time. The development toward nationalism was slow, uneven, and very mixed. Nor is it secure in its major European centers. There is a strong human proclivity, reaching back to the socialization process itself and familial care, that identifies "homeland" with home rather than nationalist abstractions, hence even the most consolidated nation-states are more divided internally than nationalist myths would have us believe. For example, the intense domestic conflicts that beleaguer the United States—notably between ethnic groups, regions, and even localities, not to speak of economic classes and interests—are testimony to the hold of specific cultural and territorial identities among populations, even those that belong to highly nationalistic "superpowers." These divisions are more strongly evident in modern Russia.

Nationalism exercises its strongest power over the popular imagination in oppositional ways when a nation is in some sense under attack. As an internally cohesive force, it has always been fragile. The success of National Socialism in "unifying" the German people has deeper roots in German history than in a sense of national allegiance, a history that is notable for the tragic decline of German social life after the Reformation and the need for compensatory mechanisms to counteract the slump that marked that bloody and enervating period.

But how did these nation-states come into existence? What role did European cities play in forming them? And what began to happen to cities—and with them, politics and citizenship—when the nation-state asserted its sovereignty over public life?

We have seen that nationalism existed only in an incipient form in the ancient world where, with few exceptions, people psychologically and culturally identified themselves with their villages, towns, cities, and immediate territorial area. Classical antiquity was mainly an era of empire building, not of nation building, and ancient imperialism took a very special form. It was patrimonial, not strictly political; its center was the royal household, not a capital in any nationalist sense. If "all roads lead to Rome," it is not because Rome was a city that commanded a high degree of national loyalty but rather because it was the hub of ancient power and order. Its centrality was derived from its administrative significance, not, as is the case with Jerusalem, its symbolic significance. Indeed, insofar as Rome was symbolic of anything, it was seen as a symbol of oppression. Crucifixion, a mode of punishment normally reserved for rebels and intransigent slaves, imparted to the cross a symbolic meaning not unlike that which was attached to the Nazi swastika during World War II. The adoption of this symbol by the Church instead of the fish, which early Christians used as an expression of plentitude and conversion, reflected the accommodation of the heavenly city to the worldly city—Rome—and of a rebellious creed to the institutions it originally opposed. Rome, once the whore of the world, to the early Christians became the holy city of the papacy and acquired a spiritual status in the very act of seducing its most intransigent rebels, a lesson in statecraft that far too many rebels have failed to learn.

Medieval Europe did not have nations at all. Loyalties were mainly localist in character, notably to one's village, parish, town, city, barons, and rather tenuously to one's monarch. Strong kings such as Henry II of England and Philip Augustus of France had an unusual amount of power for their times and, by their innovations as well as their actions, exercised a great deal of authority over their barons and clerics. More often than not, however, their heirs were weaklings and royal authority was easily subverted by feudal lords until another strong monarch surfaced and managed to centralize power for a brief period of time. Personal authority and the forcefulness of a monarch's character created the illusion of a nation-state, but it was still along patrimonial lines. The association of the monarch with the "nation," such as it was, was so close that "nationalism" in the Middle Ages rarely survived the person of the king. The wide swings in royal authority that we encounter in European feudalism stem from the tenuous basis of centralized authority as such in the ancient and medieval worlds. Centralized power was limited by the hard facts of a primitive communications system and a largely Neolithic technology. Armies and officials could move no faster than horses; roads were very poor, often almost trackless; and weaponry had barely emerged from the Bronze Age. The simple fact that obedience depended heavily upon personal ties and a reasonable modicum of fear of a resolute king resulted in a loosely hanging political system based on individual commitments and the use of brutal punitive measures. The times were cruel rather than barbarous because of the nature of statecraft in a world where authority was extremely fragile and punishment took a very harsh form in order to sustain monarchical and baronial power.

The one institution that commanded widespread allegiance beyond that which any monarch could hope to achieve was the Roman Church and the papacy. Shaped over centuries into a vast hierarchical bureaucracy, the Catholic Church eventually became the most unifying and certainly the most centralized apparatus in western Europe. It could reach into villages with an effectiveness that was the envy of any royal authority, and its command of resources had no precedent on the continent. I refer not only to its enormous wealth and landholdings but also to its spiritual power—ultimately the power of excommunication and interdiction that brought even emperors such as Frederick Barbarossa to their knees.

Yet here one encounters a paradox to which J. R. Strayer gives deserved emphasis.[1] The papacy as early as the eleventh century, during the time of Gregory VII, was an empire unto itself. More important than its centralized institutional structure is the compelling fact that it was an elaborate state: a vast apparatus not only with its black-robed bureaucracy but a system of law and a juridical machinery adequate enough to make its legal sovereignty effective. The emergence of ecclesiastical law served to open a crucial area for the development of secular law. In the great division of power that Gregory VII created between church and state, the latter essentially disentangled itself from the former. Secular law, rarely free in the past of ecclesiastical influence and priestly functionaries, could now come completely into its own. In the very act of charting and firming up its own authority, the Church created an ever-growing space for the expansion of the state and for a largely nonreligious form of statecraft that no longer required supernatural sanction and legitimation. However much monarchs were to claim divine right to support their authority, their rights were divine, not their origins— origins that even the Roman Caesars claimed. Hence, like all rights, they could be challenged even on religious grounds by the church, the barons, and later by the people. Rights are always subject to rational evaluation and legitimate defiance, a problem that monarchs such as Charles I of England and Louis XVI of France were to face before accusing regicides. That the Church, too, by claiming a legal domain of its own, was to pay a heavy penalty by being removed later on from every sphere other than its spiritual authority is a part of history that adds a dimension of irony and paradox to all assertions of power—the paradoxical reality of history's own dialectical process.

The great division between state and church, monarch and pope, secular and divine, can be taken as a symbolic expression for the many divisions that riddled medieval society and made the nation-state possible. Had medieval society been entirely unitary, it might very well have been blanketed by large, suffocating, patrimonial empires that would have drained its resources and diverted it into a relatively stagnant dead end. Ironically, it was the fragmentation of the European world that made its unification into several nation-states possible. In contrast to empires, nations are relatively self-contained; their existence, at least between the sixteenth and eighteenth centuries, was possible only because empire building by one state was blocked by the power of another. Nation building, in effect, occurred not only because monarchs accumulated sufficient power to establish a central authority in parts of western Europe but also because their ability to extend that power beyond a region was arrested by countervailing powers external to them. Charles V's efforts to create an empire in the sixteenth century failed because other states acquired enough power to successfully resist him and his son, Philip II. The successful revolt of the Netherlands and the defeat of the Spanish Armada by England ended any serious attempt to create an empire within Europe up to our own time, despite the Napoleonic wars and Germany's expansion from 1940 to 1945. Nation-states and nationalism have vitiated all efforts to create empires on the continent, at least by direct means. Russian expansion toward the west, seemingly vigorous after 1945, has visibly begun to ebb. American power in Europe, although rarely exercised in a direct manner, stands at odds with its strongly republican institutions.

This external system of "checks and balances" that medieval Europe produced on a continental level has its parallel in an internal system of "checks and balances" that existed on a domestic level. The intercity trade that provided economic underpinnings to the formation of nation-states repeatedly came up against strong localist barriers—artisan guilds, relatively self-sufficient village economies, popular hostility to trade that formed the legacy of medieval Christianity, and a very strong individual desire, particularly among craftsmen and free peasants, to live directly by their own skills and on their own property with minimal commercial intercourse—a state of mind we would call "self-sufficiency," today, which has its roots in the Greek concept of *autarkeia*. Nor can we ignore the immense hold of the *Gemeinde*, the local community, on the minds of people during that period, a sense of communal loyalty that was reinforced not only by cultural bonds but by a tradition of communal land ownership and mutual aid.

The considerable attention that recent economic historians have given to the role of foreign trade in producing a capitalistic society and the nation-state is somewhat misleading. That the earliest forms of capitalism in the ancient as well as medieval worlds were commercial and that port and riverine cities often enriched themselves enormously, promoting a domestic market for goods acquired abroad, does not remove the need for a balanced perspective toward the reach this economy had into the depths of emerging European nations. Owing to the poor condition of the roads that existed in Europe, the cost of commodities brought from abroad rose enormously as they moved inland. A cartload of hay priced at 600 denari during Diocletian's time increased by 20 denari every mile so that the cost of transporting it as little as 30 miles rendered its sale in a nearby market prohibitive. Unless a good network of waterways existed, foreign trade had a limited economic impact on the interior of a region. Generally, the commodities that passed from port and riverine cities to inland towns were luxury items or direly needed goods in limited supply. Spices, expensive cloths, skillfully crafted artifacts and weapons, and exotic foods from abroad were the usual fare that were carried from port cities into the interior of Europe. Within the heartland of the newly emerging nations, trade was often local rather than national. Intercity trade brought towns and villages together but in regional networks and local markets, not in national ones.

This "parochialism," to use a term that is distinctly pejorative today, gave inland towns a certain degree of economic strength: their relative self-sufficiency and ingrown cultures provided them with the material and spiritual fortitude to resist the overbearing effects of the nation-state, particularly during the "Age of Absolutism," when strong monarchs came to the foreground of European political life.

Decentralism remained a major obstacle—and, for generations, an abiding civic impulse—in counteracting the extension of royal power. Hence, emerging European nation-states were often checked internally as well as externally from successfully engaging in empire building. European empire building found an outlet not in the "old" continent but abroad, in so-called "new" continents such as the Americas, where Indigenous peoples living in tribal communities were easily quelled by recently discovered firearms. Needless to emphasize, European epidemics, too, played a major role in depleting aboriginal populations.

The rise of European nation-states, so unique when compared with the imperial systems of antiquity and the fragmented social structures of feudalism, can thus be explained by a special conjunction of forces. A few are simply negative: the internal resistance within emerging nations to centralism that made a highly organized patrimonial state difficult to achieve and the balance between new states that confined them to a "national" scale. Other factors, more positive in nature, consist of the fairly organic links established between municipalities primarily by local and regional trade and secondarily by the exceptionally rich rewards that port cities acquired from foreign commerce. The royal power that built the nation-state, virtually brick by brick as it were, used the internal intercity networks as political routes for extending its bureaucracies into the depths of the country just as the Church, centuries earlier, sent its missionaries and holy orders into the forests and farmlands of barbarian Europe along trails and footpaths.

Despite the enormous wealth monarchs gained from the taxes these commercial networks provided, nation building was still primarily a political phenomenon. Economic explanations to the contrary notwithstanding, the absolute monarchs of the sixteenth century who dominated the emerging era of European nationalism drained the economy as much as they fostered its development. The removal of feudal tolls on roads and riverways, the domestic tranquility provided by the "King's peace," the creation of a national currency and presumably a reliable monetary system were vitiated by the enormity of royal taxes, the ongoing wars between nation-states that placed Europe in a chronic state of siege, and the monetary instability produced by royal loans and defaults.

The role cities played in this development was very mixed; indeed, to understand it, we must know what kind of city we are talking about and what phase of nation building we are examining. Generally, the smaller, artisan-oriented towns tended to oppose the nation-state and often gained support for their opposition from the relatively free peasantry in their environs. Indeed, it could be said in a very broad way that where artisans were the majority of a free town, they elected for municipal freedom and were more predisposed to forming town confederations than larger, commercially oriented cities. Their high sense of independence and their striking defiance of authority was not to end with their opposition to royal rule; whether as master craftsmen or as skilled workers, they shared an intuitive inclination toward a vague kind of republicanism, often even democracy, that clashed with the politically more conservative merchant, banking, and professional strata in their own communities and in strongly commercial cities.

The radical role played by skilled artisans in precapitalist society and in periods of transition to the nation-state and capitalism is recently gaining the attention it deserves. This is not to say that this on-going stratum which persisted as a major social force even after the Industrial Revolution, which virtually destroyed it, shared a completely unified outlook. Between a master craftsman who enjoyed the privileges conferred by guild membership and a skilled worker who was barred from the status of a master by increasing guild exclusivity, a serious gap developed that could rend communities in violent struggles. We have seen evidence of such struggles in Flanders and Italy, where skilled as well as unskilled workers were proletarianized and reduced to exploited wage earners. But elsewhere in Europe, the traditional bridges between master craftsmen and skilled workers served effectively to close the gap that had been created by status differences; indeed, in many places guild doors remained fairly open well into the Middle Ages.

What is so noticeable about this town stratum, taken as a whole, is the fairly high educational level it attained. Artisans were often well-read, surprisingly well-informed, and intellectually innovative. Craftsmanship sharpened not only one's dexterity and aesthetic sense; it sharpened one's mind as well. The radical heretical movements within medieval Christianity consisted of large numbers of well-read artisans, not simply "millenarian" peasants, as Marxist historians such as Eric Hobsbawm would have us believe.[2] The Brethren of the Free Spirit and many revolutionary gnostic sects that challenged the church's hierarchical system were filled with artisans. Town revolts against ruling bishops, barons, and commercially minded patricians were initiated by well-informed, even well-read, artisans who could argue holy writ with disquieting acuteness. They often provided the ideological coherence for these uprisings that historians tend to impute to professional strata, not only mass support. This tradition did not die out with the Middle Ages. It continued for centuries, through the Reformation, which was by no means only a peasant war, through the era of the democratic revolutions, and into the Industrial Revolution, indeed for several generations afterward.

Skilled workers, particularly craftsmen, are remarkably self-contained individuals—the urban counterpart of the yeoman farmer who formed the basis for the democratic revolutionary era. Skill and a minimal competence, whether in the form of a personally owned shop or land holding, confers a sense of independence and self-esteem upon a person. It is out of these traits that the classical—and, to some extent, the modern—ideal of citizenship was fashioned. The lack of a respected skill and the absence of material independence makes for mobility in the pejorative sense of the term, notably, a wayward, fickle, and free-floating mob that is easily manipulated, no less by patrician demagogues than by dispossessed ones. Rebellion that lacks men and women with the moral substance produced by a sense of material independence and self-esteem—or, at least, a sense of hope that these attributes can be acquired by concerted action—tends to degenerate as easily into nihilistic counterrevolution as it does into constructive revolutionary behavior. The gap that emerged between the skilled *popolo* of Italian cities during the High Middle

Ages and the more or less lumpenized proletariat was to last for centuries and disappear only when craftsmanship was dissolved by the factory system.

European absolutism repeatedly came up against this human element in its endeavor to place its territorial realm under royal authority. That many artisans were in the service of the royal court as producers of luxury items does not tarnish the character of those who lived outside the orbit of absolutism and opposed it. These outsiders, in fact, were the great majority in Europe and generally serviced local and regional markets. By contrast, the bankers, professionals, and merchants whose incomes were derived from loans to the upper classes, bureaucratic positions, and international trade—the "big bourgeoisie," as it were—were generally royalist and more irresolute than the artisans in their localist loyalties. Absolutism provided them not only with domestic security and a huge source of revenues; banker and merchant followed in the wake of royal armies, garnering the spoils acquired by conquest, the need for arms, and the opening of new markets. Where these strata were deeply entrenched in a municipal establishment, they tended to resist but rarely rebel against royal authority. Here they might act as an inertial force against nation-building but seldom as a major obstacle. Hence they formed opportunistic alliances with rebellious artisans that they readily betrayed when things threatened to get out of hand and demands for civic autonomy threw municipalities into violent and serious opposition with the centralized state.

The "bourgeoisie" that is said to have given support to absolutism against the feudal lords consisted mainly of bankers and merchants and later, to some extent, of the new industrial bourgeoisie. Indeed, this "bourgeoisie" was strikingly conservative; it consistently supported some kind of royal power, whether absolutist in the sixteenth century or constitutional in the eighteenth. Republicanism and democratism (a very revolutionary concept until well into the nineteenth century) found its most steadfast recruits in artisans and, less reliably, the "mobility" people or rootless, statusless, and homeless "mob," whom even the English Levellers viewed warily because of their dependence on the largesse of the possessing classes.

Our picture is further complicated by the aspirations of the territorial lords, who openly rose in rebellion time and again against the monarchy—and, of course, by the peasantry that never lived on comfortable terms with the royal power, the manorial estates, or the towns. The story of the conflict between the monarchy and nobility, a struggle that reached its violent zenith in the Fronde, which literally chased the young Louis XIV out of Paris, is the stuff out of which the conventional history of European absolutism is written. It does not require emphasis here. The peasant wars in England in the late fourteenth century and in Germany in the early sixteenth were really climactic events in a chronic conflict that went on for centuries from the High Middle Ages to our own time. John Ball and Thomas Munzer were the ancestors of a host of later agrarian leaders such as Emilian Pugachev in Russia and Emiliano Zapata in Mexico. All of them shared the central goal of preserving the village community, its insular lifeways, its networks of mutual aid, its common lands and communitarian economy.

Nation building during the era of absolutism was unique precisely because it occurred within—and by the tensions it produced and fostered—a highly diversified and creative social arena for political development. The ancient world had been dominated by massive, generally immobile empires that suffocated social development with their parasitism and massive bureaucratic structures. Medieval Europe was an embryonic world—rich in promise, to be sure, but largely unformed, lacking in cohesion, indeed more particularistic than decentralistic. Absolutism and the early era of nation building stood somewhere in between. As Perez Zagorin tells us: "In spite of the expansion of absolutism and its critical collisions with rebellious subjects, royal government never became uniform or monolithic, even in the kingdoms where it suffered least limitation. Confusions of jurisdictions and rivalries and conflicts among governmental institutions abounded under its rule. Its actual power was more like an intricate mosaic of particular prerogatives, rights, and powers than a homogeneous, all-inclusive authority. The attributes of sovereignty that were ascribed to absolute kings by royalist lawyers and political philosophers, such as the French thinker Jean Bodin, were more often than not greater in theory than the powers these same kings could dispose of in practice. The monarchies of our early modern states continued to be in many regards conservative and traditionalistic, as befitting regimes ruling hierarchic societies and of centuries-long growth from ancient origins. Absolutist kings did not quarrel with social privilege, of which their own position was the supreme manifestation. They quarreled only with the privileges that resisted their authorities or claimed immunity from their government [which, one is obliged to add, seems to mean that they opposed all social privilege but their own]. In building their more centralized states, they did not sweep the stage clean with a new broom, and many older and outmoded political forms and institutions survived the reign of absolutism, like scenery left standing while a new play was performed."[3]

What Zagorin seems to reproach the absolutist monarchies for—namely their failure to "sweep the stage clean with a new broom" and achieve a more homogenized, efficient, and smoothly working social order—is what gave the era its incredible dynamism and creativity. It was precisely the rich social features of the sixteenth, seventeenth, and eighteenth centuries—the extraordinary diversity of social life and its forms, the creative tension from which so much mobility in status, cultural, and intellectual fervor emerged—that makes the era so fascinating and provocative. The absolute monarchies, caught in a complex web of ancient privileges and rights, thus became the reluctant conservators and the gnawing subversives of an accumulated wealth of old customs, autonomous jurisdictions, traditional liberties, and hierarchical claims to authority.

Within this immense warehouse of historical rights and duties, communities and individuals, institutions and interests could move with remarkable freedom and function with exciting potentiality often invoking past precedent to defend recent innovation in an old guise and play history against present to create a new future. Marx was to reprove the French revolutionaries of 1848 for dressing their rhetoric, claims, and ideals in the slogans of the French revolutionaries of 1789-94, who in turn borrowed the language and postures of the Roman Republic to legitimate then own ideals. Yet it was precisely this rich sense of historicity and variety of forms, manners, and ideas—whose power and insightfulness we will explore in the closing chapter of this book—that gave such fecundity to both French revolutions and opened such a sense of promise in Europe two centuries ago. What an individual "thinks of himself"—not simply the hidden economic laws or some unknown "spirit" that ostensibly guides history—has a very profound effect on how he or she acts as a social being and deeply influences the course of social development, Marx to the contrary notwithstanding. The richly variegated images that people had of their own time, their rights, and themselves profoundly shaped much of the history of that time. Diversity made for greater choice and a more fertile social landscape, however often precedent was invoked and reshaped to serve new ends. The tensions created by the alternate developmental pathways that emerged before communities, orders, classes, and, yes, individuals fostered a highly creative practice in the achievement of their ends.

Nor is it true, as Marx tells us, that "No social order ever perishes before all the productive forces for which there is room in it have developed."[4] I leave aside the fact that this dictum begs the very question it poses: we presuppose that the limits of a social order are defined by their "productive forces," notably technology and the various ways of organizing labor, indeed that they constitute the essence of a "social order." Actually, we have no way of knowing how far the "productive forces" could have developed in precapitalist society, nor do there seem to be any limits to the development of "productive forces" today, which means, by inference, that a market society will be with us forever.

But what is most disquieting about Marx's vision of social change is the extent to which it denies the power of speculative thought to envision a new society long before an old one becomes intolerable or is bereft of any room for development. Artisans and peasants in the era of nation-building had many sound ideas about the kind of society they wished to create. It is a mechanistic arrogance of the most extreme kind to predetermine how far communities can go—whether forward, backward, or both, in their vision of the good society and the elements that compose it—and literally dictate the material or productive limits of their vision, its "preconditions," its possibilities, and its place in a theory of history structured around various "stages" and presuppositions.

We may say, returning to Zagorin's description of absolutism, that it was not in spite of a monarch's "brutal collision with rebellious spirits," but because of this collision, that the nation-builders of Bodin's time failed to "sweep the stage clean with a new broom...." English, French, and Spanish monarchs, like their ancient counterparts who built huge empires, wanted very much to achieve "homogeneous and all-inclusive authority." They did not require the existence of local hierarchies to justify the development of a national hierarchy, nor did they need the challenges created by local privileges to justify their own claims to national privilege. To the extent that they tolerated local hierarchies and privileged strata, it was because they did not have the resources and the economic supports to replace them. Jean Bodin's justification for the all-inclusive authority of every national government over the rights claimed by lesser, more local jurisdictions is evidence of a prevailing vision among the monarchies of his day that clashed with the reality of their situation and profoundly unsettled the real world. What European monarchs did "think" of themselves—namely, all-powerful rulers of highly centralized states—deeply affected the development of European history. Their centralistic proclivities clearly influenced the thinking of men like Robespierre, France's centralizer par excellence, who also struggled to overthrow royalism as such, using Rousseau-like principles of a democracy that "forces" men to be free. Herein lies one of the great ironies of the modern era: leaders of the popular opposition to absolutism were to function in many ways like the monarchs they professed to oppose and, ultimately, they were to turn against the very popular movements that brought them to power.

It was the artisans and peasants of the autonomous towns—whose walls centralistic royalists such as Cardinal Richelieu were to demolish—not only self-governing and parochial feudal lords, who were to rise in rebellions against the monarchies of Europe. I have already examined the resistance and alternative offered by the *Comuneros* and the German peasants as examples, together with others who typified the most important efforts to arrest centralism and statist forms of nation building. The seventeenth and eighteenth centuries, no less than the sixteenth, exploded in numerous albeit more localized conflicts of the kind that produced so much civil turmoil in the sixteenth. To catalogue these smaller but highly numerous revolts, particularly those that occurred in France, would be tedious. Their overall pattern, however, is described by Zagorin very ably:

> In every case, some new exorbitancy of royal finance, a tax on taverns, for instance, or the abolition of as exemption from certain imposts, sparked the outbreak. Then the residential quarters of the poor erupted, mobs of three, five, and even ten thousand took to the streets, and the pursuit of the *gabeleurs* [tax collectors] became the order of the day. Riot and excitation, violence coupled with aspects of a popular festival and its drunken indulgence, dominated these outbreaks, as the lower orders dropped their workaday routine for the collective action that demonstrated their anger and made them feel their strength. The militant base was always the urban plebs: not made up entirely, perhaps, of what the president of the Parlement of Bordeaux called "the lowest part of the people," but consisting of artisans, journeymen, day laborers, small shopkeepers, and sometimes womenfolk, too, who participated with their men. At Bordeaux a boatman and a tavernkeeper, at Agen a master glover, a pack-saddle maker, and again a tavernkeeper, at Périgeux a physician, stood out as leaders and haranguers of the crowd. Peasants swarmed in from the outskirts to unite briefly with the revolt, as happened in Bordeaux. At Agen six thousand peasants who tried to enter were kept out by the bourgeois militia and burned the property of *gabeleurs* in the vicinity.[5]

A more detailed account would include attacks upon the town hall, the release of jailed prisoners, the killing of tax collectors and finance officials, even of wealthy individuals who were suspected of being in collusion with the state. Without the leadership provided by artisans, small proprietors, occasionally professionals, and even members of the lesser nobility, it is unlikely that these uprisings would have gone very far. Apparently their extent depended upon "the passivity, neutral or ambiguous behavior, and even partial support [of] the notables and well-to-do. The bourgeois militia of Bordeaux showed little will to act.... At Périgeux the bourgeois guard was similarly slack. Only at Agen, where the violence was particularly intense and accompanied by a definite social antagonism, did elites and authorities take concerted steps against the outbreak."

The limits of these insurrections, rural as well as urban, should be noted. Again, Zagorin comes to our aid:

> Taken as a whole, French urban rebellions were devoid of programs or ideas. In the same way as the concurrent agrarian revolts, they seemed to reflect a stasis in French society despite its exceptional amount of violence and insubordination, an inability by insurgent forces to transcend the existing state of things with any informing vision of an alternative. Although again and again the centralist offensive led to resistance, the latter failed to give rise to demands for political changes.... Urban rebellion was essentially a defensive response by the whole community. To the plebeian populace, the punishment of *gabeleurs* was a deserved act of communal justice. If such people took the largest, most active part in these insurrections, that was because they were the most sensible of fiscal oppression. But elites usually shared responsibility in some measure. Their own opposition to exactions helped to legitimize resistance, and they encouraged revolt, whether by actual instigation, covert approval and sympathy, or inaction, when disorder broke out.... But they were helpless to hold back the advance of absolutism and its centralizing grip because they had no political reforms to propose and conceived of no new institutional limits on the power of the royal state.[6]

Actually, Zagorin's account of the way these chronic town insurrections unfolded—ultimately leading to the French Revolution itself a century or so later—fails to explore certain important details that help us understand the extent to which the uprisings actually did "hold the advance of absolutism and its centralizing grip" under control. The French monarchs such as Louis XIII and Louis XIV were strong kings who were guided by even stronger advisors such as the cardinals Richelieu and Mazarin, respectively. They created the strongest armies and created the most far-reaching bureaucracies in western Europe in their day. They were ruthless centralizers and patently hostile to their own nobility, often encouraging trade and playing the "bourgeois" classes against the noble classes who were as protective of their particularistic privileges as they were of their economic ones. Western Europe in the seventeenth and eighteenth centuries was a status-oriented world. It was made up more of "orders"—status groups who enriched themselves, developed extravagant patterns of conspicuous consumption, and spoke in a mannered language according to an elaborate body of etiquette—rather than of classes who simply and unadornedly pursued their material interests. Power and social recognition were no less important than wealth; in fact they were major sources of material wealth. Personal enrichment, in turn, was often pursued precisely to gain more power and greater social recognition. Consumption in all its dazzling forms thus assumed greater significance than production, even for the bourgeoisie, which tried to elbow its way into the social hierarchies of the era. Mary Beard's account of the merchant's values in the Roman world holds as well in the era of absolutism as it does in the era of empire, a thousand years earlier. The trader, she observes, "had drained the wealth of the West for 'bones and stones' [gems and ornaments] from the East and thereby brought on currency troubles for which there was apparently no remedy.... Because he had yielded much to the landed gentleman's ideal, he tended to draw fortunes away from the business to the land, taking wealth out of circulation and energy out of business enterprise. His children, growing up in the country, were often lost to business entirely."[7]

In western Europe, especially in "classical" France—the country that so many radical and liberal thinkers were to regard as the model for "pure" feudalism and the "bourgeois-democratic" revolution—this process of moving from the status of trader to that of nobleman was especially marked and adds considerable ambiguity to what we mean by the word "bourgeois." Many "bourgeois" and noble interests coalesced. Both orders generally had a greater fear of the "mob" than of each other—and, in moments of crisis, of the monarchy. The two orders were united by bonds of marriage, land ownership, cultural tastes, values, and a desire for status and the power, political as well as economic, that it conferred.

The "bourgeois militia" of Bordeaux, Périgueux, and Agen probably included many aspiring master craftsmen, merchants, and a good sprinkling of "industrialists" who were eager to acquire noble status—even if only "nobles of the robe" (nobles who had noble status conferred upon them for service to the state or, quite crassly, for money) rather than "nobles of the sword" (hereditary nobility). They were more like medieval burghers, than capitalists in the present-day sense of the term. Indeed capitalism, conceived not simply as an economic system structured around commercial transactions or trade but, in its truly modern sense, as an accumulative system (what Marx called "expanding reproduction") in which industrial expansion became an end in itself, was a feeble thing in western Europe in the seventeenth and eighteenth centuries, more characteristically Dutch than either English or French. Hence, it would be of value to discriminate between the words "bourgeoisie" and "capitalists" rather than turn them into synonyms.

By the same token, the "lowest part of the people," whom the president of the Bordeaux *Parlement* designated as the principal activists in the urban rebellions of the 1630s, were not simply laborers, vagabonds, and beggars. In a status-oriented society, they could also be said to include master craftsmen, artisans, shopkeepers, and small-property owners. The fact that tavernkeepers, master glovers, pack-saddle makers, and even a physician "stood out as leaders and haranguers of the crowd," to use Zagorin's words, tells us something about the "crowd" they led and harangued. That nobles, too, occasionally became such "leaders" tells us something about the trans-class nature of such uprisings and the extent to which nearly all sectors of society could be united against the centralization of power by the monarchy. Most of the people hated men like Richelieu, Mazarin, and members of the king's administrative bureaucracy, despite their ritualistic expressions of affection for the king. Such cries as "Long live the king without the *gabelle!*" meant that the king, ostensibly beloved for his symbolic status as a caring father, could earn popular loyalty only if he was shorn of the all-important fiscal means for maintaining his commanding position in the country—a position that amounted to very little without the money for provisioning his army, paying his bureaucrats, and carrying on domestic wars against recalcitrant towns and nobles.

Which raises Zagorin's question of how much of a program French insurrectionaries actually needed if French society was indeed in a "stasis." Zagorin's use of the word *stasis* stands at odds with his notion that the insurrectionaries were "helpless to hold back the advance of absolutism and its centralizing grip." The "stasis" of the time consisted precisely of a lasting balance of forces between the central power and the regions, localities, orders, and classes of the time, not an ever-encroaching central power. Richelieu and Mazarin, working ruthlessly on behalf of their monarchs, tried to forge a completely centralized nation-state around the king and his bureaucracy. It is myth, in my view, to believe that any but the most naïve of the absolutist French kings trusted their nobles, although one French monarch, Louis XVI, tried to enlist his nobles in 1788 against an ever-demanding "Third Estate." They betrayed him without scruples and, in the most self-seeking manner, brought about his downfall as well as their own.

Thus Richelieu and Mazarin did not succeed, in fact. They carried centralization far in France, but not so far that it could prevent the French Revolution, which tried in its early phases to reverse the power of the king and the aristocrats alike, initially driving toward a relatively decentralized, indeed minimal, state. If centralism is to be identified with the interests of the newly emerging capitalist economy insofar as the "bourgeoisie" seeks a politically unified country, the French Revolution was categorically not a "bourgeois revolution." Nor did the French "bourgeoisie" want a "democracy" or even, necessarily, a republic. Indeed, it tried to model France on England's constitutional monarchy. The supreme irony of the French Revolution is that a movement designed to curb privilege, hierarchy, and, above all, centralized power was twisted by liberal intellectuals, "forward-looking" lawyers, and marginal strata into the most centralized state in Europe in order to defeat its counterrevolutionary enemies. In some respects, the Jacobin Republic that followed the collapse of these hopes for a constitutional monarchy and a fairly decentralized political structure was marbled by feudal concepts of government, notably, the famous "maximum" that limited the rise of placed prices and wages, state intervention in many parts of the economy and society, and a moral posture that espoused "republican virtue" as a political calling. The Jacobin Republic, in effect, was a wartime state that Napoleon was to make fairly permanent. This quixotic turn of events etched itself deeply into French national life for nearly two centuries, and centralism became a focal issue around which popular movements were to rebel throughout the nineteenth century. The climax of this popular movement was reached in May 1871 when the Commune of Paris tried to establish a confederal republic based on a Commune of communes, a struggle that ended in a bloody massacre and the ebbing of decentralistic notions of socialism in France itself.

Robespierre and Napoleon, more than Richelieu and Mazarin, were the true forgers of the nation-state in France, and it was they who provided the real "classical" example of centralized national authority in Europe.

What role did the towns and cities play in abetting this development from the sixteenth century onward? Towns and cities responded, as we have seen, in very mixed ways to the emergence of a centralized authority that tried to absorb them. Some resisted that authority with very little internal division or difference of opinion. Others, particularly where the "bourgeoisie" and elite orders prevailed, favored a well-coordinated society in which trade could prosper. They thereby abetted the work of Richelieu, Mazarin, and Napoleon, both the original article and his nephew, Louis Bonaparte, in the 1850s and 1860s. But there is a sense in which all the towns of Europe unwittingly abetted the formation of nation-states, even as they opposed it for social, political, and cultural reasons. They did this by creating the arteries and veins, as it were, for national consolidation: the elaborate, indeed historically unprecedented, road systems and nodal market towns that fostered the monetization, commercialization, and industrialization of nations—and with these possibilities, the effective centralization of the state along national lines.

I wish to advance the view that this was one of the most important achievements of Europe during the Middle Ages and well into recent times. So much emphasis has been placed by historians on Europe's development of a market economy, its relative freedom of trade, and its openness to industrial innovation—features that are certainly uniquely European—that we fail to recognize how little they would have been able to affect the continent's development without a network of roads, canals, and rivers to make such an economy possible. Apart from the trade that created market towns—towns where all goods were systematically exchanged—long-distance commerce was an elitist affair and served the well-to-do, not society at large. Local trade was necessarily very local because of the insular nature of most European communities. So, too, was political power. Rivers, canals, and maritime transportation served to link a fair percentage of Europe's towns and villages together, but by no means the majority. A European "hinterland" existed almost everywhere on the continent, such that towns and villages were almost entirely self-sufficient in the staples of everyday life, and culture remained highly insular. These "hinterlands" were largely beyond the reach of the nation-state and surprisingly impenetrable to commerce, monetization, and the "bourgeois" incubus that was to claim Europe for itself during the nineteenth century.

We can begin to sense the importance of road systems in economic development and political control today (perhaps even more than during any other part of our century) because of the effectiveness of guerrilla warfare in the Third World. The very lack of "development" and the poor road systems that have existed in Central America, mountainous portions of Asia, and the rain forests in southeast Asia have brought even nation-states aided by "superpowers" to their knees. All the "colonizers" of the past recognized the need for effective road systems. The Romans were able to hold their empire together precisely because it was networked by a highly elaborate system of roads that penetrated into the Iberian peninsula, Gaul, and portions of Britain and Germany as well as the hinterlands of the Mediterranean basin. Even quasi-tribal societies such as the Inca of Peru developed "empires" because roads—actually well-maintained pathways and footbridges—made central coordination possible. It matters little whether elaborate road networks emerged because "economic forces" created a need for them or because they made the emergence of such "economic forces" as a market economy possible. The two, in fact, may have interacted reciprocally with each other: road building stimulating commerce and commerce stimulating road building.

Judging from Rome's network, however, good roads in themselves did not necessarily foster a high volume of commerce, although they were indispensable for commercial advances. This great network of antiquity was not constructed primarily to transport merchandise. It was created mainly to move troops and imperial administrators to various parts of the empire. Even the transport of goods to supply the legionnaires was a secondary factor in road design and its routes. As R. S. Lopez tells us in his absorbing discussion of land transport, Roman troops were encouraged to live "as far as possible on the food produced in their immediate vicinity and the administrative centres were almost all situated either on the sea or on internal navigable waterways."[8] This orientation is very "un-European": it reflects the astonishing extent to which a world oriented toward agrarian values dominated its merchants and morally denigrated their sources of income. Essentially, it was the "merchant, glutted, or rather contented with his profit, [who] retires from the port to become a landed proprietor," Cicero tells us, "...[who] seems to me worthy of full praise," not the "small-scale" trader, much less, artisan, "who should be of little account."[9]

Accordingly, Roman roads were built almost defensively along mountain crests, very much like fortresses, rather than through populous valleys. Their seemingly indestructible large blocks of stone gave them a rigidity that increased their vulnerability to climatic changes, opening large surface cracks that, as Lopez tells us, "needed gangs of slaves, or continuous work by those obliged to furnish road-services to the state, and even so they were unsuitable for heavy or broad vehicles."[10] The Roman road system, indeed ancient road networks generally, were not designed to expand local commerce beyond its regional confines or open trade routes to a continental exchange of ordinary staples. Long-distance trade of luxury goods and delicacies transported by sea or along rivers was favored over a mosaic of local and regional commercial transactions that brought ordinary goods and money into remote areas of the continent.[11]

By contrast, Europe placed a high premium on the local artisan and merchant. At a time when so much emphasis is placed on the role of long-distance commerce in forming capitalism and the modern age, it is important to recognize how local producers and local markets slowly reworked the interior of the European continent, preparing it for a monetized economy and providing the goods and markets for creating a market society. This slow but historic change in the economic culture of Europe—a slow but decisive change in attitudes toward trade, production for sale, and the use of money—was the work in large part of small towns, not only busy, glamorous ports.

> Little by little a new network of roads was put into effective operation, different totally in structure and methods from the ancient one and used by quite different types of transport and haulage. The changes in transport which accompanied the commercial revolution and the outburst of economic activity in this period can themselves be described as a slow revolution. If we consider them more closely, we find that they reveal the changed political and social face of Europe. For in medieval Europe there was no large and powerful state like the Roman Empire. Slavery disappeared and in many regions, of which Italy was the first, serfdom was abolished. Finally, the dignity of the merchant and of trade won recognition, though with many reservations. Hence private initiative or local associations largely took the place of coercive government action. The routing of roads reflected the needs of commerce rather than the convenience of soldiers and civil

servants. Maximum use was made of animal labour in transport, while human labour was so far as possible economized.

Lopez goes on to say that road construction "initiative came from autonomous local bodies or independent private companies. Plurality of roads, which is always possible when the dogma of the straight line is abandoned, enabled a great number of minor centres to be included in the network."

To say, Ferdinand Braudel reminds us, that "Napoleon moved no faster than Julius Caesar" misses a number of very important points. Even if this statement were true, Napoleon had a greater variety of transportation choices than Caesar and climatic factors limited his strategic decisions far less than those of his Roman counterpart; so, too, for the European tradesman by comparison with the Roman. Even if we examine the regional network of roads created by this gradual transportation revolution, new possibilities and a sharper division of labor between the artisan and food cultivator—with its concomitant elaboration of skills, local interaction, use of money, and growing wants—all constituted the indispensable spadework for creating a sophisticated market economy.

What is important is that it was the proliferation of craftspeople, of new wares and technics, of more craft organizations or guilds, and the intensive as well as extensive development of market interactions—generally intimate, humanly scaled, and ethical—that is most characteristic of the market economy that emerged in Europe. Let me emphasize that this rich patchwork of commercial relationships enhanced human interaction, unlike the capitalistic type of market economy that was to replace it centuries later. Even as it slowly advanced over the European continent, trade on a small, local scale between closely associated communities and individuals fostered cooperation rather than competition. It tended to deepen and extend local ties rather than sever them. It was more participatory than adversarial, more moral than predatory, and politically it provided the individual producer with a stronger material base for the exercise of citizenship.[12] Finally, it tended to dissolve barriers between local communities and adjacent regions, opening the doors to new ideas, cultures, values, and the interchange of skills and technics. In time, the towns themselves acquired greater political and economic weight in the struggle to form nation-states. It may have entangled them in the crises of nation-building, but it also turned them into centers of revolutionary social transformation, an advance signaled by the *Comunero* movement in Spain and the sectional movement in France.

These markets and towns were not capitalistic. Marx has appropriately designated their economy as "simple commodity production" in which profit-making and the plowing of surplus earnings back into capital expansion are largely nonexistent. Taken in this pure form, his account of their "mode of production" is more of a Weberian "ideal type" than an actual reality. Capitalism almost certainly did exist in European towns, even predominantly in some artisan ones, and even dominated many European cities, particularly port and riverine cities. But Europe also developed for centuries by elaborating its artisan-oriented towns rather than by significantly transforming them. Viewed as a whole, the continent from the fourteenth century onward until well into the eighteenth and early nineteenth centuries seemed to have a mixed economy, which was neither predominantly feudal, capitalist, nor structured entirely around simple commodity production. Rather, it contained and combined elements of all three forms. That one of these elements seemed to be of greater significance than another in some communities and areas does not alter the mixed nature of economic life as a whole. Contrary to the widely held belief that a capitalistic economy emerged within the "womb" of feudalism and asserted itself as a "dominant" form with the attempt to build nation-states in the sixteenth century, it seems more accurate in my view to say that capitalism coexisted with feudal and simple commodity relationships. This long era, perhaps spanning more than five centuries, simply cannot be treated as "transitional" without reading back the present into the past and teleologically structuring European history around an Aristotelian "final cause" called "capitalism."

At various times and in various places, this mixed economy assumed a very balanced form—in some respects more stable in terms of everyday life than our own war-ridden century. Marxist and liberal historians have been vexed to give this long era a name that accords with the dogmatic "stages" theories of history that abound in Marx and Engels's writings. Its existence seems to be an affront to the neat categories that even radical historians impose on the rich flux of social development. Conclusions that it was either "feudal" or "capitalist," or what able writers like Paul Sweezy designate as "precapitalist commodity production" (a designation that tells us we know more about what it was not than what it really proved to be), often stand in flat contradiction to the data at our disposal.[13]

A more open-minded reading of the data and the empirical evidence at our disposal on the era—one that extended well beyond the sixteenth-century period of nation-building—will show, I submit, that western Europe existed within a force field in which feudal traditions interacted with simple commodity and capitalist forms of societal organization. Nor were feudalism, capitalism, and simple commodity production socially rigid economic systems. They were evolving cultures, differing in values, sensibilities, ways of experiencing the world and organizing it artistically and intellectually, not simply "modes of production," to use Marx's terminology again. They constituted unique social ecologies, as it were, that were patterned in highly diverse ways throughout the continent. They also changed with the passage of time in varying degrees and with varying accent or emphasis on their components, often as a result of their interaction with each other.

Space does not allow me to make a closer examination of this seemingly "transitional era," which in actuality forms a very definable and long era of European history. That it fits none of the dogmatic interpretations advanced by many modern historians who have an ideological axe to grind says more about academic trends in contemporary historiography than the richly varied texture of European life over the past thousand years: The "stasis" to which Zagorin alludes in his account of the state's development seems also to have occurred in town development during the fourteenth and fifteenth centuries, possibly even later in many parts of Europe. Although greatly increased in number, European towns seem to have entered into a demographic slack period where urban populations ceased to grow and, in some cases, diminished. Trade, too, appears to have declined and the development of new towns generally came to an end.

A number of historians have called this period the "crisis of feudalism," a terminology that may very well express a typically modern identification of "progress" with expansion. This bias tends to overlook qualitative growth and cultural elaboration for quantitative growth and cultural change. Apparently, "progress" is to be defined primarily as material growth, the increasing "domination of nature." Cultural stability and the elaboration of cultural traits, by contrast, is often conceived as evidence of stagnation," even of "crisis" and social regression. The received wisdom of historians to the contrary, is it not possible that Europe had paused during these centuries and begun to live with minimal changes in the "force field" of the era, perhaps even reaching a fairly stable balance between the three components of the mixed economy that began to develop centuries earlier? That violence was endemic at that time, as it generally was before and afterward, does not alter the stability of everyday life, especially in the small towns that contained the majority of Europe's urban population. This much is reasonably clear: by the seventeenth century, western Europe and its towns seem to have reached a historic crossroads. The continent's further "development," a term that by no means denotes "progress" in any qualitative sense, was to depend less on the growth of a centralized state and on the expansion of commerce than on technology—on the development of machines and transportation techniques that were to rework all the traditional ties that had produced such an ecologically extraordinary cultural, political, civic, and economic diversity of social and urban forms.[14]

The market society we call "capitalism"—a society that tends to reduce all citizens to mere buyers and sellers and debases all the ecologically varied social relationships produced by history to the exchange of objects called "commodities"—did not "evolve" out of a feudal era. It literally exploded into being in Europe, particularly England, during the eighteenth and especially nineteenth centuries, although it existed in the ancient world, the Middle Ages, and with growing significance in the mixed economy of the West from the fourteenth century up to the seventeenth. It is still spreading around the world—intensively in its traditional Euro-American center and extensively in the non-European world. Its forms have varied from the largely mercantile (its earliest kind) through the industrial (its more recent eighteenth- and nineteenth-century forms) to the statist, corporate, and multinational forms of our own time. It has slowly penetrated from its special spheres, such as market arenas of exchange and the production of commodities in cottages and later in factories, into domestic life itself, such as the family and neighborhood. This is a fairly recent "advance" that can be dated most strikingly from the midpoint of the twentieth century. Its invasion of the neighborhood, indeed of villages and small towns into the recesses of the domestic or familial relationships, has subverted the social bond itself and threatens to totally undermine any sense of community and ecological balance and diversity in social life.

Moreover, the newly gained dominance of the capitalistic market relationship over all other forms of production and consociation is a major source of what I have denoted as "urbanization"—the explosion of the city itself into vast urban agglomerations that threaten the very integrity of city life and citizenship. What makes the market society we call "capitalism" unique, even by contrast to its early mercantile form, is that it is an ever-expansive, accumulative, and, in this respect, a cancerous economic system whose "law of life" is to "grow or die." Capitalism in its characteristically modern and "dominant" form threatens not only to undermine every "natural economy" (to use Marx's own terms), be it small-scale agriculture, artisanship, simple exchange relationships, and the like; it threatens to undermine every dimension of "organic society," be it the kinship tie, communitarian forms of association, systems of self-governance, and localist allegiances—the sense of home and place. Owing to its metastatic invasion of every aspect of life by means of monetization and what Immanuel Wallerstein calls "commodification," it threatens the integrity of the natural world—soil, flora, fauna, and the complex ecocommunities that have made present-day life forms and relationships possible—by turning everything "natural" into an inorganic, essentially synthetic form.[15] Soil is being turned into sand, variegated landscapes into level and simplified ones, complex relationships into more primal forms such that the evolutionary clock is being turned back to a biotically earlier time when life was less varied in form and its range more limited in scope.

The effect of capitalism on the city has been nothing less than catastrophic. The commonly used term "urban cancer" can be taken literally to designate the extent to which the traditional *urbs* of the ancient world have been dissolved into a primal, ever-spreading, and destructive form that threatens to devour city and countryside alike. Growth in the special form that singles out modern capitalism from all earlier forms of economic life, including earlier forms of capitalism itself, has affected what we still persist in calling the "city" by leading to the expansion of pavements, streets, houses, and industrial, commercial, and retail structures over the entire landscape just as a cancer spreads over the body and invades its deepest recesses.

Cities, in turn, have begun to lose their form as distinctive cultural and physical entities, as humanly scaled and manageable political entities. Their functions have changed from ethical arenas with a uniquely humane, civilized form of consociation, free of all blood ties and family loyalties, into immense, overbearing, and anonymous marketplaces. They are becoming centers primarily of mass production and mass consumption, including culture as well as physically tangible objects. Indeed, culture has become objectified into commodities as have human relationships, which are increasingly being simplified and mediated by objects. The simplification of social life and the biosphere by a growth-oriented economy in which production and consumption become ends in themselves is yielding the simplification of the human psyche itself. The strong sense of individuation that marked the people of the mixed society preceding capitalism is giving way to a receptive consumer and taxpayer, a passive observer of life rather than an active participant in it, lacking in economic roots that support self-assertiveness, and community roots that foster participation in social life. Citizenship itself, conceived as a function of character formation, and politics, as part of *paideia* or the education of a social being, tend to wane into personal indifference to social problems. The decline of the citizen, more properly his or her dissolution into a being lost in a mass society—the human counterpart of the mass-produced object—is furthered by a burgeoning of structural gigantism that replaces human scale and by a growing bureaucracy that replaces all the organic sinews that held precapitalist society together.

Let it be said that this debasement of the ecological complexity of the city, of its politics, citizens, even of the individuals who people its streets and structures, is of very recent origin. It did not really begin in a manorial society, with its barons and serfs, food cultivators and artisans, and all the "orders" we denote as feudal. Nor did it follow from those grossly misnamed revolutions, the "bourgeois-democratic" ones of England, America, and France, that ostensibly catapulted capital into political control of a society it presumably "controlled" economically during earlier generations. Rather, this development began to appear with technical innovations that made possible both the mass manufacture of cheap commodities and, what is crucially important, their increasingly rapid transportation into the deepest recesses of Western Europe, inexpensive and highly competitive with the products of local artisans who serviced their localities for centuries. It need hardly be emphasized that this development depended enormously for its success on the opening of colonial markets abroad: the Americas, Africa, and particularly Asia, the area where the English crown found its richest jewel, notably India.

It was the extraordinary combination of technical advances with the existence of a highly variegated society, relatively free of the cultural constraints to trade that prevailed in antiquity, that gave economic ascendancy to the capitalistic component of the mixed economy over all its other components. Neither wealth from the Americas nor the large monetary resources accumulated by port cities from long-distance trade fully explains the rise of industrial capitalism—a form of capitalism that more than any other penetrated into the very inner life of Europe.[16] Had the wealth acquired from the "New World" been a decisive factor in creating industrial capitalism,

Spain rather than England should have become its center, for it was Spanish *conquistadores* who initially plundered the Aztec and Inca empires and brought their precious metals to Europe. The very wealth these "empires" provided for the ascendant nation-state in Spain served to weaken town life in the Iberian peninsula and provide the means for absolute monarchs to embark on an archaic program of continental empire building that eventually ruined Spanish cities and the countryside alike.

Nor did long-distance trade provide the most important sources for capitalizing industrial development. Rather it fostered consumption more than production, the dissolute lifeway that makes for a diet of luxuries instead of the parsimonious habits that steer investment into new means of production. Indeed, too much state centralization and too much commerce, despite the wealth they initially generated, ultimately led to excessive expenditures for territorial expansion and high living by elite groups in all the orders of a courtly society. That nation-building, increased centralization, or, more properly, national consolidation prepared the way for industrial capitalism by opening more "hinterlands" to trade is patently clear. So, too, did increases in the population of dispossessed, propertyless hands, whether as a result of land enclosures or normal demographic growth—hands that were available for a factory system that had yet to appear on the economic horizon. Europe, in effect, was more open than any part of the world to the expansion of its capitalist component along industrial lines. This was especially true of England, a country in which a remarkable balance had emerged between a particularly bourgeois-minded aristocracy and monarchy on the one hand and a very open economy on the other, peopled by a tight-fisted, god-fearing Puritan yeomanry and structured around towns with decaying guild systems and cottagers in the countryside who were already engaged in a domestic form of textile mass production. Whether the new industrial capitalists who were destined to totally reshape the country emerged from the merchants involved in the sale of raw wool to cottage weavers and its sale in the towns, or whether it came from well-to-do artisans, yeomen, aristocrats who used sheep to create a new form of pastoral agribusiness, or merchants who were inclined to siphon their wealth into production—possibly even all of them combined at one time or another—need not concern us here. What pushed the capitalist component of this mixed economy into a nation that could regard itself as the "workshop of the world" in the nineteenth century was a series of inventions that made the factory system and the distribution of its wares possible.

Nor need we be concerned with whether the needs of a "rising bourgeoisie" produced the Industrial Revolution or the Industrial Revolution gave rise to the "bourgeoisie," which in any case was always a presence in all the major cities of Europe. Factories, in fact, had begun to appear in eighteenth- and even seventeenth-century England long before an industrial technology had emerged. Wherever the "bourgeoisie" entered into the productive sphere rather than the commercial, it tried to bring labor together and rationalize output even with tools, hence a strictly technological interpretation of the *rise* of industrial capitalism would be greatly misleading. My concern here is how industrial capitalism managed to gain ascendancy over *other* forms of production, including commercial capitalism, and alter *all* social relationships that encountered its power. Waterwheels had preceded the steam engine as a prime mover, and work sheds organized around simple tools had preceded mechanized factories. But without the inventions that introduced the Industrial Revolution in the late eighteenth and early nineteenth centuries, it is doubtful that industrial capitalism could have impacted so powerfully on Europe and ultimately on the entire world.

England, perhaps more so than other countries in the western European orbit, seems to have been an ideal terrain for technological innovation. Its society was highly mobile socially and extremely fluid territorially. The civil war of the Cromwell period—a conflict that brought armed Puritans to power and diversified the country's system of government—had shattered the few remaining feudal inhibitions to change. Even the English aristocracy had been readied for a gainful capitalistic economy by the wealth it acquired from the sale of wool. We shall have occasion to note that landed squires and industrial capitalists were not without serious differences over the distribution of manpower in the country's economy, nor did the rural patronal system disappear completely. But it can be said that England, more than any other European country at the time, had begun to acclimate itself to industrialization owing to the relative weakness of absolutism, which might have drained the nation's wealth to fulfill imperial ambitions and the growth of a money-minded yeomanry and a highly avaricious merchant class.

One must consult Max Weber's *The Protestant Ethic and the Spirit of Capitalism* to understand how much gain became a quasi-religious "calling" among England's merchants and its well-to-do yeomanry and artisans; how much they were readied psychologically for the plundering of their own communities and country. The Spanish *conquistadores* had ruthlessly sacked the Indian "empires" of the Aztecs and Inca in order to enrich themselves, not to "save" their "souls." They were maniacally involved in acts of destruction that often verged on self-destruction with only indirect effects upon the texture of Spanish society. The English Puritans, by contrast, dourly plundered their own country in a systematic and lasting manner to accumulate capital. Their behavior seemed to be guided by a moral imperative that made greed an end in itself, an imperative that was to transform the lifeways of the precapitalist society from which they had derived. Long after Puritanism had passed away as a movement, the joyless spirit that the Puritans brought to England then lived on, largely bereft of the few moral constraints that religion generally imposed on the acquisitive spirit engendered by the Reformation.

Within a span of some two generations, England was transformed on a scale unprecedented in the history of western Europe. Friedrich Engels's *The Condition of the Working-Class in England*, a period piece based on personal observations in 1844, could justly call the changes introduced by the new industrial inventions—principally in textiles, metallurgy, and transportation—a historic change of unprecedented proportions. The rapidity of the transformation is what makes these changes so startling in a domain of human endeavor—technology—which had developed over centuries at a slow, piecemeal pace. The social and cultural ramifications of this technological revolution were nothing less than monumental. "Sixty, eighty years ago, England was a country like every other," Engels tells us, "with small towns, few and simple industries, and a thin but *proportionately* large agricultural population. Today [1844] it is a country like *no* other, with a capital of two and a half million inhabitants; with vast manufacturing cities; with an industry that supplies the world, and produces almost everything by means of the most complex machinery; with an industrious, intelligent, dense population, of which two thirds are employed in trade and commerce, and composed of classes wholly different, forming, in fact, with other customs and other needs, a different nation from the England of those days."[17]

Engels's words are meant as much to praise the "revolutionary" work of the English industrial capitalist as they are intended to startle the German readers, who first encountered his book. The Enlightenment's uncritical commitment to technical progress completely infects the author's generalizations, however much he is bitterly critical of the misery inflicted by industrial capitalism on the new proletariat. Accordingly, we learn that the preexisting cottage weavers, who were to be devastated economically and culturally by the new textile machines invented around the middle of the eighteenth century, "vegetated throughout a passably comfortable existence, leading a righteous and peaceful life in all piety and probity; and their material position was far better than that of their successors. They did not have to overwork; they did no more than they chose to do, and yet earned what they needed. They had leisure for healthful work in garden or field, work which, in itself, was recreation for them, and they could take part besides in the recreations and games of their neighbours, and all games—bowling, cricket, football, etc., contributed to their physical health and vigour. They were, for the most part, strong, well-built people, in whose physique little or no difference than that of their peasant neighbours was discoverable. Their children grew up in the fresh country air, and, if they could help their parents at work, it was only occasionally; while of eight or twelve hours work for them [introduced, later, by the factory system] there was no question."[18]

The mixed messages in this passage reflect the characteristic tension created among the "progressives" of the era between the claims of an abstract sense of "history" and an existential sense of humanity that was to rend apart major socialists such as Engels and ultimately, owing to his commitment to industrial progress as a "precondition" for a free society, justify the most horrible barbarities of Stalin as the "modernizer" of Russia. The "unquestioning humility" of the English cottage weavers and yeomen, "their silent vegetation...which, cosily romantic as it was, was nevertheless not worthy of human beings," to use Engels's pejorative remarks, is doubtful on its own terms, particularly from the standpoint of our own time when the ebullient concept of "progress" that prevailed over the past two centuries is so much in doubt.[1920] But placed in a broader historical context, it is hard to square this denigration of the cottage weavers and lesser yeomanry with their vigorous role in the English Civil War of the 1640s and their resistance to the introduction of machinery—indeed, the role they played in the Chartist movement, the most radical cause, culturally as well as economically, that England has seen since the Puritan revolution.

The explosive nature of the Industrial Revolution and the emergence of industrial capitalism can best be understood by noting the limited time span in which the new technology surfaced and swept over England—a period so brief that it is almost difficult to call it an "era," as so many historians do. James Watt's steam engine, which freed the factory system from its reliance on waterpower, was invented by 1769; James Hargreaves's spinning jenny, which enormously increased the output of yarn, by 1764; Richard Arkwright's throstle, a sophistication of the spinning machine calculated for the use of a mechanical prime mover, by 1767; Samuel Crompton's mule, by 1785, followed around the same time by Arkwright's carding machine and preparatory frames; finally, Edmund Cartwright's workable power loom by 1785. All of these inventions totally revolutionized the production of textiles as it had been pursued for thousands of years and catapulted English cottons, woolens, linens, even lace and silken goods (as a result of another group of machines invented in the first decade of the nineteenth century) into a largely agrarian domestic and world market. The opening of extensive coalfields and iron mines, owing in large part to the use of coke, and the staggering enlargement of smelting furnaces on a scale fifty times bigger than earlier, more customary sizes, paved the way for transportation changes that were to open the most remote areas of the country to industrialization and "commodification."

The men who invented the textile machines and prime movers that powered them were not members of the aristocratic class and, by no means accidentally within the context of modern English history, included a sprinkling of very pious individuals, even clergymen. Watt was an instrument maker; Hargreaves, an engineer; Arkwright, a barber; Crompton, an ordinary textile worker; and Cartwright, a minister, who went on to work with Robert Fulton in the development of the steamship. English society had opened itself, more than most European societies, to advancement by individuals from lower social ranks. Moreover, many of the inventors whose work introduced the Industrial Age were to become major capitalists in their right. Only a few exhibited the breadth of spirit that characterized Robert Owen, a textile manufacturer who demonstrated at New Lanark that relatively humane conditions for workers did not conflict with the acquisition of high profits and later placed his life and his fortune in the service of the rising labor movement.

The transportation revolution that paralleled the Industrial Revolution is of monumental proportions in the development of modern capitalism itself, a revolution that a century and a half later was to turn into a "communications revolution" whose effects have yet to be fully grasped. During the first forty years of the nineteenth century, Scotland, much of which had been an untamed "hinterland" only two generations earlier, was interlaced by nearly a thousand miles of good roads and an even larger number of bridges. This was largely "the work of private enterprise," Engels tells us, "the State having done very little in this direction."[21] Indeed, by 1844, England, apart from Scotland, had well over two thousand miles of canals, substantially in excess of its navigable river mileage. Finally, the early railroad lines that linked Manchester to Liverpool, were rapidly extended across the entire industrial midlands, sweeping in Lancaster, Sheffield, Leeds, Newcastle, and, of course, London further to the south.

English industrial and commercial towns, too, exploded in size and population. Within the first thirty or forty years of the nineteenth century, Birmingham increased its inhabitants from 73,000 to 200,000; Sheffield from 46,000 to 110,000; Halifax from 63,000 to 110,000; Leeds from 53,000 to 123,000. Such astonishing rates of growth among comparatively small communities in so short a time had never been seen before in European history. The congestion and diseases they produced have been fully chronicled in the novels as well as the histories of the period. For the first time, apart from a few cosmopolitan centers in antiquity and perhaps some Flemish and Italian towns in medieval times, we begin to see the authentic devouring of cities, not only the countryside, by urbanization. Birmingham, Sheffield, Halifax, and Leeds, among many others in England, were beginning to lose their identity as cities in any classical meaning of the term. Their factories, sprawling out into the countryside, were phasing into urban agglomerations, sweeping in villages and small towns that had existed with relatively little change for centuries. William Wylde's panoramic painting of Manchester in 1852 reveals a forest of smokestacks and a dense collection of squalid dwellings, a new, formless, and repellent urban world that industrial capital began to introduce into the world—first in England, then in France, Belgium, Germany, and, of course, across the Atlantic where the grimmest images of Pittsburgh as a steel-making center seem interchangeable with Wylde's picture of Manchester as a textile center generations earlier.

Yet, what is very important about this period of rampant industrialization is the fact that the more commercial and administrative cities of England and even the industrial towns that blighted the island's landscape were to develop a communal character of their own. Growing industry, commerce, and "commodification" did not seep completely into the neighborhood life of the new cities, nor did it totally destroy the conditions for the regeneration of domestic life. The buffeting that towns and cities of the nineteenth century took from industrialization, however disastrous its initial effects on traditional lifeways, did not destroy the inherently villagelike subcultures of workers and middle-class people who were only a generation or two removed from a more rural culture. Like the ethnic groups that entered the New World through New York City throughout much of the nineteenth and early twentieth centuries, displacement was followed by resettlement and recommunalization, even in the most desperately poor slums of the overpopulated cities of Europe and America. The pub in the industrial cities of England, the café in France, and the beer hall in Germany, no less than the various community centers around which the ethnic ghettos formed in New York and other American cities, provided foci for a distinctly working-class culture, largely artisan in its outlook, class-oriented in its politics, and knitted together by mutual self-help groups.

This recolonization of community life was greatly abetted by the organized labor movement in all its different forms. Socialist clubs, trade-union centers, local cooperatives, mutual-aid societies, and educational groups created a public space that included classes in reading, writing, literature, and history. The socialist clubs and union centers provided libraries, periodicals, lectures, and discussion groups to "elevate" worker consciousness as well as mobilize them for political and economic ends. Picnics, athletic activities, outdoor forays into the countryside served to add a very intimate dimension to purely educational projects. The *casas del pueblo* established by Spanish socialists and the *centros obreros* established by the Spanish anarchists, which existed up to the late 1930s, are reminders of the vigorous development of community life even in the most depressed areas of Europe—indeed, of an "underground" culture that always paralleled the received culture of the elite orders and classes.

There was always a plebeian cultural domain at the base of society, even in the most dismal and squalid parts of ancient, medieval, or modern cities, that was beyond the reach of the conventional culture and the state apparatus. No economy or state had the technical means, until very recently, to freely infiltrate this domain and dissolve it for a lasting period of time. Left to itself, the "underground" world of the oppressed remained a breeding ground for rebels and conspirators against the prevailing authority. No less urban in character than agrarian, it also remained a school for a grassroots politics that, by definition, involved groups of ordinary people, even in sizable communities, in a plebeian political sphere and often brought them into outright rebellion. This "underground" school created new political forms and new citizens to deal with changing social conditions. Even after the great boulevards of Baron Haussmann ripped into the plebeian *quartiers* of Paris, opening the city to artillery fire and cavalry charges against barricades, the sizable neighborhood pockets left behind retained an imperturbable rebellious vitality that finally culminated in the Paris Commune. Few of Europe's major cities were spared crowd actions and uprisings in the nineteenth century, indeed well into the first half of the twentieth.

As industrial capitalism spread out from England into western Europe and America, the initial destabilization it produced as a result of urbanization and mechanization was followed by a regeneration of popular culture along new patterns that also included the integration of old ones. Just as the French village was reproduced as *quartiers* in French cities and the Spanish *pueblo* as *barrios* in Spanish cities, so the Jewish *shtetl*, the "Little Italys," and "Little Irelands" were reproduced in altered form but with much of their cultural flavor, personal intimacies, and traditional values in world cities such as New York. Even the industrial cities replicated on a local basis the specific cultural origins of their variegated populations and regions.

The elaboration of the Industrial Revolution into a textile economy in England, a luxury goods economy in France, an electrochemical economy in Germany (itself a second industrial revolution in the last half of the nineteenth century), and an automotive economy in America did not eliminate this "underground" communal world. Which is not to say that any of the countries cited above did not also expand economically and technologically into the key industries that distinguished each from the other. But the textile manufacturing that catapulted England into the world market and even helped to create it; the fashioning of luxury goods that gave France, especially Paris, its distinction as the artistic center of the world; and the mass production of electrochemical commodities and automobiles by Germany and the United States, respectively, which finally began to transform the world into its present form—all seemed to carry over into the popular culture and give it specific national features. We find a fixidity in England along industrial lines that appears in its conventional and underground cultures as well: a strong system of orders, open enough to add titles to the names of its gifted common-ers but consciously hierarchical in its cultural and social structure with a tendency to retain social status and elaborate it along stable lines. The intellectuality of France, expressed by its aesthetic empha-sis in industrial production, carried over into the popular culture as a discursive politics that nourished an ideologically innovative work-ing class. Germany, which seems to have straddled the hierarchical culture of England and the intellectual culture of France, gave rise to a philosophically inclined popular culture that took pride in its academic spokesmen as surely as English workers were endeared to noblemen who often spoke on their behalf. The pragmatism and dynamism of American culture, with its strong republican values, carried into the popular culture as various images of the "American Dream," partly utopian, partly materialistic, in which the automobile symbolized enhanced sexuality together with freedom of movement, both socially upward and territorially outward. This symbolism was to shape American architecture and city planning by producing the first skyscrapers and the most sprawling, homogeneous "cities" of the twentieth century.

Yet every class culture was always a community culture, indeed a civic culture—a fact that links the period of the Industrial Revolution and its urban forms with precapitalist cultures of the past. This con-tinuity has been largely overlooked by contemporary socialists and sociologists. While the factory and mill formed the first line of the class struggle in the last century, a struggle that in no way should be confused with the class war that is supposed to yield working-class insurrections, its lines of supply reached back into the neighborhood and towns where workers lived and often mingled with middle-class people, farmers, and intellectuals. Wage earners had human faces, not merely mystified proletarian faces, and functioned no less as human beings than class beings. Accordingly, they were fathers and mothers, brothers and sisters, sons and daughters, citizens and neighbors, not only "factory hands." Their concerns included issues such as war and peace, environmental dislocation, educational op-portunities, the beauty of their surroundings as well as its ugliness, and in times of international conflict, a heavy dose of jingoism and nationalism—indeed a vast host of problems and concerns that were broadly human, not only class-oriented and rooted in wages and working conditions.

This communal dimension of the industrial era is of tremendous importance in understanding how class conflicts often spilled over beyond economic issues into broadly social, even utopian, concerns. Indeed as long as the market did not dissolve the communal dimension of industrialism, there was a highly diversified, cooperative, and innovative domain of social and political life to which the proletariat could retreat after working hours; it was a domain that retained a vital continuity with precapitalist lifeways and values. This partly municipal, partly domestic domain formed a strong countervailing force to the impact of an industrial economy and the nation-state. Here, workers mingled with a great variety of individuals, particularly artisans, intellectuals, and farmers who brought their produce into the towns. In a purely human fashion that revealed all the facets of their personalities, they developed a sense of shared, active citizenship. This communal or municipal citizenship kept political life alive even in highly centralized and bureaucratized nation-states. It would be difficult to understand not only the radical uprisings of the nineteenth century but also the twentieth—particularly the series of urban and agrarian uprisings that culminated in the Spanish Civil War—without keeping this communal dimension of the "class struggle" clearly in mind. Every class movement from the late eighteenth to the early twentieth centuries was also a civic movement, a product of neighborhood, town, and village consociation, not only the factory, farm, and office. It was not until a technology developed that could make deep, perhaps decisive, inroads into this "underground" municipal domain that politics and citizenship were faced with the total "commodification" of society, the supremacy of statecraft, and the subversion of the city's ecological diversity and creativity.

The older generation of our time is still too close to the present and the younger too far from the past to realize the extent to which communal, precapitalist cultural traditions permeated the first half of the twentieth century. Nor do we fully realize how rapidly these traditions are being simplified and the extent to which they are disappearing in the second half of the century, largely as a result of sweeping technological and cultural developments that occurred during the Second World War and its aftermath. Science systematically applied to technics has revealed the deepest secrets of matter and life with nucleonics and bioengineering. We are vividly aware of the possibilities opened by cybernetics, space travel, and communications systems for public surveillance, not to speak of mass "mind-bending" with techniques that have made conventional advertising methods seem childish and means of detection that make fingerprints seem primitive.

The psychological implications of these chilling advances are as significant as their cultural and institutional ones. The military armamentarium they have produced is as disempowering to the individual as it is frightening to the public. The result is that the ego itself tends to become passive, disembodied, and introverted in the face of a technological and bureaucratic gigantism unprecedented, indeed unimaginable, in earlier human history. Public life, already buffeted by techniques for engineering public consent, tends to dissolve into private life, a form of mere survivalism that can easily take highly sinister forms. Citizenship, in turn, tends to retreat into an increasingly trivialized egoism in which banal conversation and venal pursuits replace searching discourse and rebellious demands. Even more than ordinary individuals, the people of mind and art who formed for ages the troubled conscience of past societies and were deeply engaged in public life as the principal catalysts for achieving the good society—in short, those socially involved thinkers whom the Russians of the nineteenth century called the *intelligentsia*—retreat into the confines of the academy and are transformed into mere intellectuals. Research tends to replace creative speculation; libraries tend to replace the cafés that nourished a Diderot; classrooms replace the squares that brought a Desmoulins to the surface of public life; the academic conference and its journals replace the clubs and presses that provided the public forum for a Thomas Paine.[22]

Industrial capitalism in the nineteenth century created a market economy in which production and exchange still formed a layer of social life, not a substitute for society. The "cash nexus" that Marx and Engels regarded as the core of the new economy in *The Communist Manifesto*—culturally as well as commercially—was evidence more of their foresight than a description of the factual conditions of their times. Important barriers stood between the penetration of the "cash nexus" created by the factory and by commerce into the community and domestic life, where the oppressed in all their different forms and even the agrarian squirearchy seethed with hatred of the new economy. The Speenhamland Law that English country squires passed in the 1790s to provide the material means for keeping the rural population out of the emerging factory system, reveals the stronghold that precapitalist patronal traditions had on a society in the very midst of the Industrial Revolution.

These traditions persisted in England well into the twentieth century, long after the Speenhamland Law was repealed, often in highly overcompensatory forms that exaggerated the smallest status distinctions within the aristocracy, middle classes, and even the working class, hence the "multiclass" nature of English society long after industrial capitalism had settled into the island. One can extend these social barriers to every country of Europe, however much they differed in form and from one region of the same country to another. Even republican America fell victim to an appetite for titles, as Henry James so clearly perceived, and to internal status distinctions: native-born versus immigrant, Yankee versus everyone else, gentile versus Jew, and, above all, white versus Black. A "feudalistic" mentality with its patronal condescension resisted the "cash nexus" that Marx and Engels, like many others, imputed to the culture of their day.

What is unique about the post-World War II era, is that the market *economy* to which industrial capitalism gave a firm and dominant productive base is turning into a market *society*. The "cash nexus" has become an all-embracing commodity nexus in which consumption, not only production, has become an end in itself. Objects now mediate nearly all the social relationships that once had a living and creative flesh-and-blood character. Popular culture, which meant the disdained culture produced from below by the people, has now become "pop culture," the synthetic "culture" generated from above by the mass manufacture of books, sounds, pictures, movies, and television. Human relationships are increasingly defined in terms of material things and expressed in a commercial vernacular—be it marriage as an "investment," child rearing as a "job," life as a "balance sheet," ideals as things one "buys," and community as a "business." The notion that towns and cities should be "managed" by "entrepreneurs" as though the civic bond has no dimensions to it other than its "efficiency" in coordinating "services" and its capacity for generating "revenues" permeates the language not only of bankers, merchants, politicians, and retailers but even liberal and—most tellingly—socialist public officials. Social life, so conceived and lived in the form of marriage, child rearing, education, friendship, conviviality, entertainment, culture, and values, reveals a degree of commodity penetration and commercialism unprecedented even in the halcyon days of "free trade" and capitalist production.

Which is not to say that the period following the Second World War did not have its preludes in the pre-war era. One only need look to the assembly line as a portent of what was soon to come. But there was much that the pre-war era left untouched. The Depression had given a bad name to capitalism and placed its legitimacy in grave doubt. This loss of faith stimulated a sweeping drift toward cultural and social movements that challenged the system's integrity and claims to meet human needs. Agrarian life still had a strong base in American and French society, even as depression foreclosures reduced its economic significance. As late as 1929, twenty-five percent of the employed labor force in the United States owned or worked labor-intensive farms that had changed very little from the last century. Approximately three-fourths of American agricultural produce was sold at home. Although tractors were beginning to appear in growing numbers, horses were still in wide use and kerosene was still a major fuel for illuminating farm homes. Culturally, agrarian values were very much alive in America. They were embodied in the high proportion of immigrants from the villages of eastern Europe, Italy, and Ireland, and in the yeoman farmers who were being scattered throughout the continent by the dust storms that swept over middle America.

The Depression, in fact, witnessed a remarkable regeneration of precapitalist agrarian values. The end of the First World War had seen a cultural mystification of urbanism and substantial migrations of young people from rural areas to cities. Songs such as "How Can You Keep Them Down on the Farm After They've Seen 'Par-ee!'" typified a glorification of urban lifestyles in the 1920s. The rural "hick," once conceived as the source of America's sturdy republican values and the "pioneer" of continental settlement, had become a caricature in the cartoons, literature, and cinema of the twenties, a "country-cousin" bumpkin who would do well to conceal his identity in the "big city." This remarkable change in the historical imagery of Americans reflected the ascendancy of the industrial city and the modern age over the small town and the Jeffersonian age—in short, the cosmopolitan capitalistic era over the parochial precapitalist era.

The Depression decade reversed this imagery dramatically. For millions of unemployed or economically uncertain Americans, capitalism was patently not a working system. For all the hoopla about the decline of socialism as a movement from its high point in the period immediately preceding the First World War, it is worth recalling that Norman Thomas, the presumed heir of Eugene V. Debs, received 900,000 votes in 1932 as the Socialist Party's presidential candidate, a figure that comes very close to Debs's highest vote some fifteen years earlier. The Depression reversed the migration from farm to city, at least until the dust storms of later years drove many farmers to other parts of the country. Rural culture also began to thrive in the mass media of the decade, particularly as an evocation of community values, mutual aid, and compassion for less fortunate people. The extraordinary degree of social vitality that marked the 1930s stands in marked contrast to the socially devitalizing selfishness of the 1920s, which celebrated Calvin Coolidge's maxim: "America's business is business"—and, one might reasonably add, let morality go to the devil. Social justice, not personal venality, rings through the thirties as the motif of the radical left and radical right. Indeed, for much of the political spectrum, the business ethic was relegated to damnation together with the values of the twenties, which to many Americans had brought on the crises of the thirties.

Characteristically, the cinematic foppery of the twenties and early thirties began to give way to movies that focused on such themes as poverty, unionism, justice, and the pristine values of agrarian lifeways. Even where capitalists are shown in a benign light, such as in *The Devil and Miss Jones*, businessmen were seen as converts to the new humanitarianism and the high sense of public responsibility that marked the decade. Others, like *Mr. Deeds Goes to Town* and *The Grapes of Wrath*, unabashedly lauded precapitalist rural values and vilified the rich, hypostasizing the face-to-face simplicity and high moral sense of the small town. The rural community was set ethically against the spiritual poverty and cynicism of the city. A cultural populism swept over America after 1935 that has no parallel with any time before the 1890s, one that gave to community, public service, political idealism, and social justice a premium that makes the decade unique in twentieth-century American history.

To a leftist of the 1930s, it seemed that industrial capitalism had reached its economic, social, and cultural limits; that from an ascending society, it had begun to descend historically, just as feudalism did centuries earlier. The Depression held on doggedly up to the Second World War; indeed, in the judgement of some economists, the closing years of the thirties appeared to be drifting into a new economic crisis comparable to or worse than the one that had darkened the decade's opening years. Technological advance, saddled by the inertial weight of an increasingly monopolistic economy, seemed to have come to a virtual end. The monographs of the Temporary National Economic Committee of the late thirties, established to assess the growth and effects of business concentration in the United States, were filled with citations of technological stagnation, indeed the suppression of technical innovations that made for higher productivity and lower prices. That these innovations, seen as a source of economic instability and oligarchic control, had been generally arrested was cited by the literature of the time as evidence of a "decline" of capitalism itself and as proof of the Marxian thesis that "bourgeois social relations" stood in a "historical contradiction" to the "mode of production." Capitalism, it was argued by liberals and leftists alike, was incapable of furthering the "productive forces" of society and could no longer play a "historically progressive" role.

Yet there were also harbingers of significant tendencies counteractive to the radical analyses that marked the decade. Rural electrification was to basically alter the traditional agrarian lifeways that the Depression decade celebrated. It brought the radio into the home and exposed a socially pristine population to the mass media. The spate of public works that marked the Roosevelt era opened roads into the remaining hinterlands of the continent and the automobile culture improved existing roads for engine-powered vehicles. The great federal highway system of a later generation was to give the *coup de grâce* to regional isolation and network the entire country with a remarkable road system that all but obliterated centuries-old cultural variations and customs. Distance and time were being overcome by the internal combustion engine, whether by rail, road, or air.

Finally, the need for state intervention into a failing "free enterprise" economy vastly strengthened the nation-state and enlarged its bureaucratic apparatus. Rural life, however much it was celebrated culturally, was being locked into urban life and the city increasingly placed in a position subordinate to the nation-state. Business, too, was becoming concentrated and centralized. By 1929, nearly half of American industrial wealth was controlled by 200 large corporations and by 1932, despite the ravages of the Depression—or perhaps because of them—600 corporations controlled nearly two-thirds of the country's industrial economy. Politically and economically, power was to concentrate and centralize into fewer hands, linked together by an endless number of bureaus, bureaucrats, and shared interests that were to prepare the way for the "technobureaucratic" nation-state so characteristic of our own time.

The Second World War, while holding the values of the thirties somewhat in place during the first half of the forties, foisted another Industrial Revolution on western society whose scope, as we are now beginning to see, rivals the shift of humanity from a hunting-gathering society to an agricultural one. By 1946, the technical basis had been created for a world that men and women of the thirties could never have anticipated. Electronics, nuclear power, giant strides in biochemical research, and the sophistication of rocket propulsion, not to speak of less visible but monumental advances in physics, chemistry, and biology that were to yield revolutionary applications of scientific knowledge to industry, were to totally change the economic landscape. Political change for "total war" had already given rise in Europe to a new phenomenon in the west—the totalitarian state—beside which royal absolutism seems in retrospect like a feeble thing. The population dislocations created by military mobilization as well as war refugees, followed by the expansion of migration of industry to the western United States, and the pouring of billions into European reconstruction, vastly transformed the urban landscape. Los Angeles, a "small" city in the thirties by present-day standards, became the "paradigm" for unlimited, unplanned, and sprawling urbanization with its individual anomie, and a predatory form of capitalistic enterprise that overshadowed the "Roaring Twenties," with its naïve imagery of personal greed and vice. The fifties introduced a more anonymous and faceless phenomenon: corporate greed and commercialized vice, the marketing of managerialism and suburban self-indulgence (material as well as sexual) as a new way of life and a new set of values.

Cities were to lose not only their territorial form; they were to lose their cultural integrity and uniqueness. *The Man in the Gray Flannel Suit*, particularly as embodied in the sartorial conventions of Gregory Peck, became the avatar for the new managerial bureaucratism, despite the movie's tribute to suburban family life. Social justice, idealism, and agrarian values of community gave way to privatization, self-indulgence, and suburban cookouts. The desiderata of a corporate age, its synthetic and highly personalized culture, was a sanitized and socially vacuous world, held together not by the "underground" culture of past centuries but by an "above-ground" culture engineered by government and corporations. A mass society, notable for its despiritualized and amoral version of "possessive individualism" had emerged from the debris of the thirties, a mass society structured around television networks, bureaucratic agencies, and, above all, commodities. The market society had begun to come into its own, destructuring the city, the domestic world, the psyche, and ultimately the natural world. "Man," brought into collision with "man" in a highly competitive world mediated by commodities, was finally brought into collision with nature on a scale unprecedented in history.

Urbanization may be taken as the symbol and the reality of this historic ecological disarray of the modern landscape—indeed this reversal of the evolutionary clock—from organic to synthetic human and biotic relationships. Nor can this disarray be sorted out and pieced together on the summits of social life. Historically, the power of preindustrial lifeways to survive and preserve a moral sense of social life depended on the existence of an "underground" culture in towns, neighborhoods, and cities—even during the stormiest periods of nation-building, industrialization, and "commodification"—to countervail the power of the market and the official culture. They depended, in effect, upon the power of a popular culture to resist the elite culture and, in no small part, on the inability of the elite culture to successfully penetrate the popular one. The crisis of our time as expressed in the decline of political life and citizenship, of community and individuality in the classical sense of these concepts, stems from the invasion and colonization of culture's subterranean domain by a highly metastatic capitalistic technics and its commodities. Objectified and fragmented, the social bond itself faces dissolution. What we call the "grassroots" of society is turning into straw, and its soil—the locality or municipality—is turning into sand. Whatever evidence of "fertility" and "life" these roots exhibit seems to be the result of toxic chemicals—mass media, bureaucratic sinews, and managerial controls—the nation-state and corporations seem to be pouring into its bedrock, just as agribusiness generates "food" out of the chemically saturated sponge we call "soil" today.

Redemption must come from below, from the municipal level where the "underground" culture of past times flourished, not from an "above" that constitutes the very source of the problems we face. The "long march through the institutions," which sixties radicals such as Rudi Dutschke advocated in Germany and Herbert Marcuse evoked in America more than a decade ago, has demonstrated that it is the institutions that ultimately absorb modern rebels, not the rebels who capture the institutions.

Urbanization, insofar as it still permits the city to retain any identity, alters not only its form but its function as a civilizing arena for humanity. The "stranger" who could once find a home in the city and a role in its political life as a citizen reverts to his or her status as an "alien." Ghettoization moves hand-in-hand with urban sprawl, the dissolution of the self hand-in-hand with the city's dissolution. Bureaucracy fills the vacuum created by the absence of an active community life. Critically viewed, the social ecology of urbanization is the compelling story of the destructuring of all social life from home to community, the erosion of heterogeneity, interaction, and civic creativity. The return of the stranger expresses the change from the "lonely crowd," explored by David Reisman decades ago in his book of same name—indeed, to the fearful crowd of today, a crowd in which everyone moves guardedly through the modern city with a sense of subdued dread toward surrounding people and unfamiliar neighborhoods.

As urbanization spreads, so too does the state machinery needed to administer it. Whatever its form, the nation-state increasingly approximates a totalitarian state and the privatized individual, riddled by egoism and fear, increasingly becomes more like one commodity among many rather than a consumer. Ultimately, society threatens to become as inorganic as an ecosystem that has been divested of its flora and fauna. When totalitarianism eventually does emerge, it is likely that the coherent self, so direly needed to resist it, will have undergone such erosion that it may well lack the psychological resources to recognize the danger, much less oppose it.

Chapter Eight
The New Municipal Agenda

From the moment nomadic hunter-gatherers began to settle down into stable villages, they introduced radically new changes that went beyond the turn from food gathering to food cultivation. Where villages became towns, human beings began to slowly "detribalize" themselves and create those civic institutions we associate with civilization. The blood tie, the male-female divisions in functions, and status groups based on age that formed the sinews of tribalism were slowly absorbed into an entirely new social dispensation: the city. Cities were structured primarily around residence, a vocational division of labor, and a variety of "orders" and classes, some of which were united by economic interests but others by power and prestige. The biological facts of blood, sex, and age, in effect, slowly phased into the social facts of propinquity, vocation, wealth, and privilege.

Out of this sweeping historical transformation, new ways of ordering life began to emerge. The biological realm of life, seemingly "natural" in its origins, became what we normally regard as the social: the realm in which people organized to meet their material needs, to reproduce as well as produce, to relate with each other as individuals and family groups, to fraternize in a variety of personal associations and degrees of intimacy. The town and city, even the village, provided a new political realm of life, human-made in origin, in which individuals related to each other as citizens in order to manage their communities and cope with its civic affairs. Until comparatively recent times, "biological" and political realms overlapped significantly, so that political elites such as aristocracies legitimated their authority over towns and cities by highly tribalistic claims of ancestry and genealogy. Until the Emperor Caracalla made all free men in the Roman Empire citizens of the state, citizenship was largely an ancestral privilege. The rights it conferred to citizens in the management of a community were marked by various degrees of ethnic exclusivity.

But the city, initially the Hellenic *polis*, created a new social dispensation: minimally, a territorial space in which the "stranger" or "alien" could reside with a reasonable degree of protection—later even a measure of participation—that few tribal communities allowed to outsiders, however hospitably they were treated. Two new civilizatory categories—politics and citizenship—began to absorb the largely biosocial forms that had held early familial and tribal lifeways together.

The professionalization of power and violence that we associate with the state came much later. Indeed, in its national form it is a fairly recent phenomenon. The state and the practice of statecraft have no authentic basis in community life—by which I mean that, until recently, the acquisition of power by professionals encountered the greatest difficulty in legitimating itself. Tribal institutions can easily be understood because human beings are, after all, natural organisms. They must feed themselves, acquire their means of subsistence, reproduce, live in safety, and given their evolution as primates, engage in some kind of communal intercourse. In contradistinction to other animals, however, they organize these activities into mutable *institutions* and make them operationally systematic—hopefully rational, predictable, and ideologically legitimate. Animals, including the genetically programmed "social insects," do not have institutions, however habituated and predictable their behavior may be; in short, they do not have consciously formed ways of ordering their communities that are continually subject to historical change.

It is important to distinguish the social in humanity from the political and, still further, the political from the statist. We have created a terrible muddle by confusing the three and thereby legitimating one by mingling it with the other. This confusion has serious consequences for the present and future. We have lost our sense of what it means to be political by assigning political functions and prerogatives to "politicians," actually to a select, often elite, group of people who practice a form of institutional manipulation called statecraft. Inasmuch as many politicians are viewed with a certain measure of contempt, we have degraded the concept of politics— once a participatory dimension of societal life and the activity of an entire community—by confusing it with statecraft, a distinctly power-oriented activity. Accordingly, we have lost our sense of what it means to be a citizen, a status we increasingly accept as mere "constituents" or "taxpayers" who are the passive recipients of the goods and services provided to us by an all-powerful state and our elected representatives.

Ideologically, we tend to justify this historical degradation of our status as political beings by invoking the nation as the basic and most elemental unit of social life, an entity that is itself of very recent origin. That nations are actually made up of cities, towns, and villages on which our wellbeing, culture, and security ultimately depend has begun to slip from our consciousness. It remains a lasting contribution of Jane Jacobs to have demonstrated in a very compelling way that our economic well-being depends on cities, not on nation-states. Indeed, that while nations may be "political and military entities it doesn't necessarily follow from this that they are also the basic, salient entities of economic life or that they are particularly useful for probing the mysteries of economic structure, the reasons for rise and decline of wealth...We can't avoid seeing, too, that among all the various types of economies, cities are unique in their abilities to shape and reshape the economies of other settlements, including those far removed from them geographically."[1] We may leave aside Jacobs's conventional use of the word *politics*, her imagery of the economy as a market, indeed, as a capitalistic economy, and her strongly economistic interpretation of cities primarily as centers of production and exchange. Her argument invites a debate about the superfluity of the nation-state that has been too long neglected, and her examples can be ignored only for the most dubious ideological reasons.

A close reading of history has demonstrated that the state—and in our own time, the nation-state—is not only the repository of agents and institutions that have made a mockery of politics; it shows, in fact, that these state agents and institutions have degraded the individual as a public being, as a citizen who plays a participatory role in the operations of his or her community. In this respect, the nation-state has impeded the development of much that is uniquely human in human beings, disempowering the individual and rendering him or her a warped and one-sided being.

It is equally demonstrable that the state—and again, the nation-state—parasitizes the community, denuding it of its resources and its potential for development. It does this partly by draining the community of its material and spiritual resources; partly, too, by steadily divesting it of its power, indeed of its legitimate right, to shape its own destiny. Despite recent rhetoric to the contrary, nothing has seemed more challenging to the state than demands for local self-management and civic liberty. Decentralization, a term that is often abused these days for the most cynical ends of statecraft, is not merely rich in geographic, territorial, and political values; it is eminently a spiritual and cultural value that links the re-empowerment of the community with the re-empowerment of the individual.

Municipal freedom, in short, is the basis for political freedom, and political freedom is the basis for individual freedom. For centuries, the city was the public sphere for politics and citizenship and in many areas the principal source of resistance to the encroachment of the nation-state. In its acts of defiance it often delayed the development of the nation-state and created remarkable forms of association to counteract the state's encroachment upon municipal freedom and individual liberties.

The argument in support of the nation-state today is almost entirely logistical and administrative. Social life, we are normally told, is too "complex" to allow for municipal freedom and participatory citizenship. This argument does not stand up very well against historical and contemporary evidence. Fascinating examples can be found of economic and political coordination within and between communities that render statecraft and the nation-state utterly superfluous.

Aside from Hellenic, Italian, medieval German, and Castilian endeavors, one of the most lasting are the Swiss communes whose practice of using local resources without turning them into private property was an object of fascination for generations to visitors in the mountain areas of central Europe. These communes and the confederations they formed for their common welfare and safety seem to have absorbed almost tribalistic forms of intimacy into their practices, even attitudes that viewed sharing according to need rather than work into their ways of dealing with material goods. According to some accounts, land, which was often open to use by all who needed it, included streams, clay pits, quarries, or, in H. Mooseburger's words, "the entire region with every and all its products and fruits."[2] Let me emphasize that this was a municipal form of "ownership," about which we shall have more to say later, not the nationalized forms advocated by statist socialists, the "economic democracy" advanced by many liberals, or systems of "workers' control" demanded by orthodox anarchists—all of which, I may add, involve some degree of state involvement or a particularistic and potentially competitive body of interests within the community.

The Gray Leagues (*Graubünden*), the source of the Swiss referendum and the town meetings of its 222 communes, have to be singled out as the most libertarian of all. Until Napoleon forced it into the Swiss Confederation, the *Freistaat der Drei Bunde* (literally, the Free State of the Three Leagues) and specifically the *Graubünden* itself, the league that gives its name to the confederation of the three leagues that composed it, was to exist for nearly three centuries (1524–1800) and place its distinctly decentralistic imprint on Switzerland as a whole. This Free State "was not merely democratic," observes Benjamin Barber in his informative account of the league and its standards of community freedom. The fact is that "it was democratic in a particular way not easily accounted for by the conceptual perspective of Anglo-American thought ... Graubünden's experience with democracy has been inseparable from its experience with community, and as a model of integral community, no region of Switzerland can equal it."³

Barber's conclusions cannot be emphasized too strongly. What they concretely indicate is that the ultimate source of sovereignty reposed in the commune—the village, town, or city—whose assent or opposition to a course of action was achieved by referendum. The confederal system that united the communes had the right to deal with foreign affairs and little more. Beyond this sphere, confederal bodies were concerned mainly with preventing their component leagues from making foreign alliances on their own. Issues such as war and peace were decided directly by the communes themselves. "The only instrument of the central government was a three-man commission (*Haupter*) made up of the heads of each of the leagues that, with the assistance of an elective assembly (*Beytag*), prepared the referendum and executed the will of the communes," Barber tells us. "In the new structure power was an inverse function of level of organization. The central 'federal' government had almost none, the regional communes had a great deal."⁴ Whatever its capacity to deal with the problems that confronted the Free State, the people were provided "with several centuries of real thoughtful independence and a measure of autonomous self-government rare in Germanic Europe."

The challenges faced by the Free State and its component leagues over these centuries require a lengthy study of the kind provided by Barber. The problem of dealing with foreign intervention; the incorporation of sizable towns and the expansion of older ones into cities; the disparities in status, wealth, and power that developed, not to speak of local parochialism at one extreme and cosmopolitan modernity at the other, were never fully resolved. But they were kept in remarkable balance for most of the Free State's history. Even after Napoleon had reduced the Free State to a canton in the more centralized Swiss Confederation, Barber notes, the peasant still turned the harshness of his sparse land "into a discipline of individuality, a teacher of autonomy. He held the intuitive conviction that his rural mountain life, his uncomplicated involvement in a pastoral economy that left him considerable leisure time, was inextricably bound up with his independence and his freedom."⁵

By no means should this strong sense of individuality be mistaken for the "individualism" associated with traditional "natural law" ideologues and the modern-day proprietary emphasis on egoism. The hardships inflicted on Alpine dwellers, locked in a glacial land of heavy snowfalls, avalanches, and floods, placed a high premium on "collective labor and common decision making. Thus as the hardness of life molded a man's sense of autonomy, it also compelled him to cooperation and collective action." As Herman Weilenmann has put it, to the villagers of the *Graubünden* freedom involved "not individual emancipation from his obligations to the whole, but the right to bind himself by his own choice."[6] Neither the individual nor the collective, in effect, claimed sovereignty over each other, but rather they formed a complementary relationship that supported each other.

These abiding notions are difficult for modern Euro-Americans to accept nowadays. Yet they are deeply rooted in the American tradition itself. I refer to the New England town-meeting tradition from which so many of the authentic libertarian aspects of the "American Dream" derive, rather than to the "cowboy" tradition that presumably "tamed the West" and often reduced it to a spawning ground of sheer avarice. Historians increasingly note the localist and communalist motives that drove Puritan settlers to New England. Religious persecution by Charles I was only one of several conditions that the Puritans found intolerable under the Stuart kings. "If Charles I is remembered at all today, it is as an ineffectual monarch who lost his head on the chopping block," observes T. H. Breen, whose work on New England Puritan institutions is perhaps one of the best that has been published in recent years. "During the first years of his reign, however, he brought considerable energy to his position. He instituted or tried to institute far-reaching changes in civil, ecclesiastical, and military affairs. These unprecedented, often arbitrary policies disrupted the fabric of local society, and they were a major preoccupation of the men and women who moved to Massachusetts Bay. One cannot fully understand the institutional decisions that the colonists

made in America unless one realizes how gravely Stuart centralization threatened established patterns of daily life in England's local
communities."[7]

The men and women who formed the New England townships
of the seventeenth century sought not only to restore Christianity
to a "pure," ecclesiastically untainted, biblical form; they also sought
to restore society itself to a pristine, egalitarian, and devoutly communalist pattern, presumably paralleling the ethical and social covenants that appear in "Acts" and the quasi-tribal democracy of the
Hebrew Bedouins and their compacts. A shared assumption seemed
to exist that the English town corresponded to the Hebrew tribal
community and that its restoration was a necessary move toward
social as well as religious purification. Every community was conceived as an ethical compact, not simply a form of association for
personal and collective survival—a notion decidedly antithetical to
the Hobbesian and Lockean principles that enter into liberal conceptions of republicanism as we know them today. Men and women,
in the Puritan world view, formed communities to achieve a "good
society" in the moral, not merely the material, sense of the term: a
society marked by virtue as defined in Christian precept. Individual
and community, in this sense, were no less inseparable among the
early colonists of New England than they were among the villagers of
the Swiss *Graubünd*.

The stormy declamations of angry prophets such as Amos, whom
Ernst Bloch so aptly called a "barn burner," sear the Puritan soul
and explain the institutional development of Puritan communities.
It was Charles I, with his "ill-advised attempt to increase his authority by attacking local English institutions," who appears as the
Moloch in this drama, and the Puritan divines as the prophets who
denounced royal encroachments on the "liberties of Englishmen," to
use the language of the day. Despite the diversity of their origins,
ranging from "populous commercial centers such as London and
Norwich" to "isolated rural communities ... most had been affected
in some personal way by the king's aggressive efforts to extend his
civil and ecclesiastical authority.... The experience of having to resist
Stuart centralization, a resistance that pitted small congregations

against meddling bishops, incorporated boroughs and guilds against grasping courtiers, local trainbands against demanding deputy lieutenants, and almost everyone in the realm against the collectors of unconstitutional revenues, shaped the New Englanders' ideas about civil, ecclesiastical, and military polity. The settlers departed from England determined to maintain their local attachments against outside interference, and to a large extent the Congregational churches and self-contained towns of Massachusetts Bay stood as visible evidence of the founders' decision to preserve in America what had been threatened in the mother country."[8]

In view of the grim reputation that the Puritan towns acquired as dour "theocracies" and parochial, self-righteous, dogmatic nests of intolerance, we must provide, with more nuance, a picture of their variety, roundedness, and militancy—not simply as they existed at any given moment but as they evolved, eventually to become centers of social rebellion, civic freedom, and collective liberty. Nor can we ignore the strong-minded yeomanry they produced, the high sense of individuality and citizenship they developed. These personal and social traits persisted in New England for some three centuries—not as mere historical ephemera that pass like wispy clouds across the social horizon but as an established democratic legacy. Deeply ingrained in the American tradition, this legacy has always acted as a force countervailing the egoistic sensibility fostered by reactionary nationalists and liberal centralists in other parts of America.

Let us quickly rid ourselves of the idea that this Yankee yeomanry was "tight-fisted," "commercially oriented," "grasping," and emotionally guarded. This imagery mistakes Boston, a commercial and acquisitive port, for New England as a whole. The port cities, in fact, often stood at odds with the many small townships that were networked together in the interior of the New England colonies and states. These colonies and their interiors were unique inasmuch as they were peppered by vital towns, in contrast to other regions of English-speaking America, which were generally marked by highly dispersed settlements and individual farmsteads or large plantations. Two facts emerge that deserve emphasis. Firstly, Boston and other port cities in New England were no different in their outlook and interests from Baltimore or Charleston; hence they were fairly atypical of the Yankee spirit of the region. Indeed, they were no different as commercial ports from port cities elsewhere in the world. Secondly, the fact that small towns could be planted so firmly, indeed, doggedly in New England's thin glaciated soil was itself an act of ethical defiance, for the region is a poor agricultural area that offers no congenial home for flourishing communities.

The towns and townships that emerged on the rocky soils of Massachusetts, Connecticut, Rhode Island, and particularly New Hampshire, Maine, and Vermont were a challenging moral statement of firm intentions to live in a very definite way, a structural expression of what constitutes a virtuous life, not merely a bountiful one. Whereas Boston eventually visualized itself as the "New World's" counterpart of London, many New England townships saw themselves as biblical communities, united by Old Testament covenants that placed a high regard on essentials over frivolities, on fairness and mutuality in relationships, on egalitarianism in status, on self-sufficiency in the development of needs and their satisfaction, and on elaboration of communitarian values over fetishization of change. In short, they were committed to a moral economy and society, not to lifeways premised on the market and gain.

We must look at these lifeways closely if we are to understand the political institutions and economic relationships that expressed them and were designed to reinforce them. However much these moral lifeways varied, owing, in part, to their ties with the commercially-oriented port cities, they were remarkably self-sufficient. I speak here of a world in which the yeomanry constituted 70 percent of the agrarian population. Crops were cultivated mainly for survival rather than trade. As one yeoman wrote in the 1770s, a farm gave "me and my whole family a good living on the produce of it. Nothing to wear, eat, or drink was purchased as my farm provided all."[9] That surplus crops were used to purchase manufactured goods such as iron, nails, glass, weapons, gunpowder, and medicine does not alter the marginality of commerce for most townships. Land utilization reflected the domestic nature of the economy. It was very sparsely used despite the large amount that was available for cultivation. Although yeomen often had 50 or more acres of land, generally only a fifth to a tenth was actually put to pasture and food cultivation, and usually only for the family and its livestock. Women were engaged in caring for the home, child rearing, making home-spun clothing, and performing other family and farm chores; men carried out the equally arduous tasks of farming, woodcutting, and construction. Diversified crop cultivation is evidence of efforts to meet domestic more than market needs.

If anything, this yeomanry seems to have viewed commerce disdainfully, indeed as parasitic and demeaning to a productive and ethical way of life. In the words of the Bostonian George Richards Minot, writing in the 1780s, yeoman-landed property "had always been held in higher esteem and more valuable nature than any personal estate." Indeed, its "possession ... seems to be of greater gratification to the pride and independence of men."[10] This form of agriculture, divested of acquisitive and commercial ends, provided the New England farmer with a sense of personal and ethical independence that made him eschew the word "farmer" or "peasant" for "yeoman" or "husbandman"—an example of the way in which men think of themselves that was to profoundly affect their behavior and the course of history.

Yet as David P. Szatmary points out, "a feeling of independence did not necessarily lead to individualism. Although priding themselves on their autonomy, yeomen lived in a community-oriented culture. To ease their backbreaking work during planting and harvesting, they asked family and friends for help. The independent status of yeomen, then, resulted neither from self-sufficiency nor a basically competitive system but led, rather, to cooperative, community-oriented interchanges."[11] More concretely, this communal orientation brought women directly into food cultivation and enhanced their social status, all talk of Puritan "patriarchy" to the contrary notwithstanding. The enormous respect—and troubled concern—that made dissenters such as Anne Hutchinson the center of religious and political controversy in the early years of the colony attest to the forwardness of women in this society and their enormous strength of character.

Barter and the sharing of resources reinforced neighborliness and fostered a warmly hospitable openness to people, even to newcomers, as the Marquis de Chastellux was to observe in his tour of New England during the early 1780s. The all-encompassing egalitarianism that pervaded this yeoman world included even agricultural laborers, a mere ten percent of the rural workforce, who usually owned some land in their own right or were given land in reward for their services. New England did not really have a rural proletariat. Indeed, disparities in wealth were too insignificant to give rise to a stable class society.

Out of this world, which had roots in Britain as well, emerged the town meeting: a direct democracy in which local and, in times of social unrest, broadly political issues were fervently debated and resolved. Here, in contrast to Swiss democracy, attempts to trace the New England town meeting back to Germanic tribal traditions may well be examples of historical overkill. The assured source of the town meeting lies very much at hand. Puritans were mainly Congregationalists, a form of Protestantism that denies by definition the need for any ecclesiastical hierarchy or centralizing body. In this libertarian conception of Christianity, all powers of religious interpretation belong to the local congregation, which is united to others of the same kind by the presence of Christ. The church, in effect, is a spiritual metaphor rather than a rigid institution. This radically decentralistic interpretation of the community as a self-governing congregation can easily extend outward into the civil world in the form of an equally self-governing political body, the town meeting. The periodic meeting of the entire male population of a community in order to govern its own affairs is a logical outcome of Puritan religious belief and forms of organization.

Property and income restrictions on the right to participate in town meetings were not taken too seriously. Initially, disparities in land ownership and the payment of taxes could have excluded only a small number of residents in most New England towns. In the more rural areas, these restrictions counted for virtually nothing. Later, by the 1760s, when colonial unrest led to outright revolution, town meetings were so notoriously open that even newly arrived or transient residents could participate in them. Historians who emphasize a link between property qualifications and New England's franchise system have proven themselves to be sticklers for regulations that went unenforced two centuries ago or were essentially nonexistent.

The reaction to Charles I's attempts to place localities under centralized control and weaken local militias heightened the sensitivity of the New England towns to their independence and right to bear arms. It is a contemporary libel regarding the role of an armed people to invoke the insecurities of the American frontier as an excuse to ban arms in the more "stable" and presumably more "secure" society of our own day. By the late eighteenth century, the New England states were largely free of Indian attacks. The prevalent notion of a covenanted township and a covenanted militia had very little to do with personal safety. The yeomen of New England were not so much fearful of Indian forays against their persons as they were concerned with statist forays against their liberties. In one of the most radical state constitutions to be adopted during the American Revolution, Vermont yeomen, gathering at Windsor in July 1777, not only abolished slavery and property qualifications for the franchise; they also avowed that "as standing armies, in the time of peace, are dangerous to liberty, they ought not to be kept up." Accordingly, "the people have a right to bear arms for a defense of themselves and the State," a right that explicitly goes far beyond the reticent wording of the Second Amendment to the United States Constitution.

These sentiments reflected the yeomanry's state of mind two hundred years ago. They had nothing to do with the "gunslinging" and bullying machismo of the nineteenth-century West. Rather, they expressed the sound conviction that the state cannot be trusted to claim a monopoly of violence over its citizenry and the full custody of public freedom. Democracy, if it was to mean anything, presupposed the active involvement of the citizen in developing a participatory politics, public security, and the direct face-to-face resolution of community problems. Lacking the means and instruments, particularly the weapons, to enforce its decisions, such a citizenry in the eyes of the yeomanry would be reduced to a mere instrument of the state, whose legitimacy in their eyes was extremely dubious.

The extent to which the town was the living repository of institutions such as the town meeting and the militia is indicated by Robert A. Gross's description of the New England yeoman's outlook. "When the eighteenth-century Yankee reflected on government," Gross observes, "he thought first of his town. Through town meetings, he elected his officials, voted his taxes, and provided for the well-ordering of community affairs. The main business of the town concerned roads and bridges, schools, and the poor—the staples of local government even today. But the colonial New England town claimed authority over anything that happened within its borders. It hired a minister to preach in the town-built meetinghouse and compelled attendance at his sermons. It controlled public uses of private property, from the location of slaughterhouses and tanneries to the quality of bread sold at market. And it gave equal care to the moral conduct of its inhabitants.... No issue was in theory exempt from a town's action, even if in practice the provincial government occasionally intervened in local disputes and told the inhabitants how to run their lives."[12]

Not surprisingly, the town meeting in even more secular form swept out of New England during the Revolution and was to extend as far south as Charleston, South Carolina. With the end of the Revolution itself, municipal "counterrevolutions" (to use the language of the historians) essentially pushed it back to the region of its origin and replaced municipal assemblies with mayors and aldermanic councils. New England and a number of towns bordering the region tenaciously, even defiantly, held on to their democratic municipal institutions, at least in the smaller towns and villages.

The "Founding Fathers" who fashioned the national constitution of 1787 created a fairly centralized republic. But they were also obliged to tolerate a basically confederal, face-to-face municipal democracy within their instrument of government and a fairly radical Bill of Rights that had been foisted upon them by a restless yeomanry. The United States Constitution, in effect, embodies a precarious compromise between demands for a municipal democracy and a centralized nation-state. Running at cross purposes throughout the document and the quasi-legal Declaration of Independence are agrarian, precapitalist commitments to freedom, a participatory politics, and an involved citizenry on the one hand, and a distinctly capitalistic imagery of acquisitive individualism favored by the rising commercial elements in the port cities and inland market towns of the new nation.

There is no way we can return to this agrarian world with its essentially Neolithic technology, its physical rigors, and the unavoidable insecurities of a premodern economy, any more than we can return to the simple religious belief systems and rustic values that shaped the New England yeoman's way of life. Nor should we try to do so. Our technology can achieve far more than theirs, and our culture, at least for the present, is more secular and perhaps more universalistic than theirs. But we can admire and learn from the early American town's communal virtues and democratic spirit, which provide us with a rewarding image of civic commitment and citizenship.

The people who settled New England regarded themselves as "Englishmen" during colonial times and fought the crown to retain "the liberties of Englishmen," which they, like the Levellers in the Puritan Revolution, believed the monarchy had violated. Many historical ties unite the American revolutionaries with the British, ties that are partly, at least, reflected in civic movements in both countries. To begin with, New England's villages were patterned very closely on the communities from which the Puritan settlers had emigrated. Indeed, many New England villagers had been neighbors in Britain and traveled to America collectively and reestablished their towns in much the same form and often with the same name as the communities they left behind. Thus they arrived with distinctly British radical traditions, influenced by the egalitarian Congregationalists, or as they were commonly called in seventeenth-century England, Independents—that is to say, the same religious radicals who were to foment the English Revolution of the 1640s and provide the troops for Cromwell's New Model Army.

Indeed, to an extent that is rarely appreciated, the American Revolution marked the completion of the English Revolution, which regressed shortly after Cromwell's death back to monarchical rather than republican institutions. In his definitive study of the most radical republicans of the 1640s and 1650s, the Levellers, H. N. Brailsford has shrewdly observed that their political vision of "an agreement of the people acceptable to the general will" did not disappear with Cromwell's interregnum and the restoration of the monarchy. "It crossed the Atlantic ... and bore ripe fruit. Defeated in Europe, the English Revolution found its triumph and its culmination in America."[13]

The radical movements that finally gave rise to the Puritan Revolution of the 1640s in Britain, particularly the Levellers—who nearly gained control of Cromwell's army—were largely based in cities and towns. The demands of the more revolutionary Puritans, who rallied around Cromwell and drifted increasingly into the Leveller movement, were buoyed by a religious spirit of "Independency," free of hierarchical control and "popish" impositions. It requires no *tour de force* of the imagination to realize that this same sense of "Independency" could, under circumstances of social unrest, easily be transferred to the political sphere—which, as we have seen, is precisely what the transplanted Englishmen of Massachusetts did during their settlement of the New World.

The New England settlers may also have been influenced by the traditional English parish's "open" vestry assemblies, initially fairly democratic, which were summoned by the ringing of local church bells. These practices in Britain eventually became fairly oligarchical, or "closed," as landed gentry and churchmen exercised increasing control over the assemblies. But the tradition of the "open" vestry assemblies may have existed among Puritan radicals and been carried to America by them. In any case, New England town meetings moved in a strongly democratic direction as the American Revolution approached. They became more open, admitting new residents to a town with few restrictions, and reducing the powers of the town selectmen who ran town affairs in between assemblies.

Unlike the American system, English town government is complicated by a long history of development dating back to Roman times. Town governments were anything but uniform: they incorporated conflicting jurisdictions created by feudal society and an emerging craft society, as well as many differences in local traditions and power structures.

To be sure, William the Conqueror, crossing the Channel with his Norman barons and knights in 1066, tried basically to remake his new kingdom by imposing a continental feudal system that was structured around a strong monarchy, a policy that was pursued by his greatest descendant, Henry II. But later kings, with few exceptions, proved incapable of following their forceful predecessors. Henry's son John was obliged by his barons to sign the famous Magna Carta, a "bill of rights" primarily for the English nobility, which restricted the powers of the crown. Other monarchs, occupied with costly wars in France and unruly barons at home, went deeply into debt and initiated a policy of selling royal charters to the cities in order to gain funds, particularly in the form of municipal tax collections. In time, these cities—towns by modern civic definitions—were to be controlled by craft and merchant guilds. And like so many guild-controlled towns, after a period of internal democracy for all master craftsmen, they tended, like the "open" parish vestries, to develop into oligarchies, in which the most established and wealthiest masters—who, of course, had the material means to give time to community affairs—exercised explicit or de facto control over civic life.

Nevertheless, a strong tradition of civic independence existed in Britain, especially in relation to the crown. Although the chartered towns varied considerably in the types of oligarchies and limited "democracies" that existed from one region to another, they firmly resisted the attempts by the Stuart kings in the seventeenth century to subordinate them to an absolute and arbitrary monarchy. Indeed, the efforts of James I and particularly his ill-fated successor, Charles I, to establish the same kind of centralized monarchy that was emerging on the Continent contributed significantly to what many British historians discreetly call the "English Civil War."

Albeit not for the first time, the invocation of "the liberties of Englishmen"—which the American colonists were to invoke against the British crown a century later—became a rallying cry of rebellious parliamentarians, Independents, town oligarchs, merchants, craftsmen, yeomen, squires, and the like. Not even Cromwell, perhaps one of the strongest of Britain's centralizers, and his military Protectorate, managed to subvert this independent civic spirit. Indeed, it was to live on with all its vibrancy well into the period of the Industrial Revolution, which brought not only manufacturing to the country but a degree of civic instability unprecedented in British history—and with it, reform movements to weaken the control of urban oligarchies and rural squirearchies. The social disruption produced by industrial capitalism reduced many British towns and cities to sheer chaos. Small hamlets were rapidly turned into squalid factory towns, while once-attractive towns expanded into sizable cities filled with displaced rural folk who provided the "hands" for rapidly expanding textile factories. The older, incestuous landed and guild oligarchies of an earlier era were patently incapable of dealing with the burgeoning towns and their infrastructural problems. Moreover, the newly emerging industrial capitalists and proletariat, both increasingly at odds with each other as well as with the old civic oligarchs, began to demand a strong voice in city affairs. Not only the town guilds but the traditional squirearchy in the countryside were fading into genteel senility.

By the 1830s, a new bourgeois reform movement and the frightening Chartist movement of the British working class forced Parliament to enact a series of laws significantly altering the structure of civic life. The Municipal Corporations Act of 1835 extended the franchise to more taxpayers in 178 towns (the traditional City of London, with its long tradition of guild management and its powerful financial interests was exempted) and established more representative forms of local government. As the labor movement became increasingly domesticated over the course of the century, Parliament extended the eligible electorate still further, embracing the urban poor and the country towns, until, by the end of the century, universal male franchise was enacted into law. Cities were divided into wards, often corresponding to distinctive neighborhoods; local councils were established on a village, town, city, and borough or county basis, and present-day forms of civic government more or less came into existence.

A brief comparison of American and British systems of local government, today, reveals several important similarities. Both systems legally regard their cities and towns as "creatures" (to use American constitutional jargon) of the state. They can exercise no rights that are not conferred upon them either by state governments in the United States or by Parliament in Britain.

Theoretically, at least, local civic policies are supposed to be formulated by elected town and city councils. In fact, however, an increasing number of American towns and cities are administered by appointed "city managers," who often have more power than city councils. In Britain, town and city councils have lost power to quangos (an acronym for "quasi-autonomous nongovernment organizations"), composed of civil servants and local businessmen, that may be established either by the national government or—let it be noted—by local councils. Quangos enjoy considerable autonomy and are generally beyond the control of the electorate. In both countries, city managers, quangos, and a variety of "authorities" (American) and "joint boards" (British) have diminished the power of elected bodies and attempted to absorb policy-making into administrative agencies.

Both countries have also seen a steady decline in the powers of local government. In the United States, state governments have steadily reduced the power of town and city councils by taking over control of municipal taxing power and local educational systems. The British Parliament, in turn, has weakened the powers that belonged to regional and local councils, partly by creating boards to control housing, transportation, and the like. In both countries, too, conservative governments have been privatizing a growing number of public services, such as educational institutions, public transportation, and, in the United States, even prisons.

A common myth I have encountered in my European lectures is that local control is a uniquely American phenomenon. In fact such goals have been advanced in recent times in France, Spain, Italy, the Nordic countries, and especially so in Britain, where socialist and anarchist notions of municipal democracy have been as strong, if not stronger, than in the United States.

British notions of "municipal socialism" can be traced back to the Fabian Society, founded in 1883, which gained such eminent adherents as George Bernard Shaw, Beatrice and Sidney Webb, Annie Besant, and a number of early founders of what became the Labour Party. In contrast to the revolutionary Marxists and anarchists of the late nineteenth and early twentieth centuries, the Fabians sought to achieve socialism gradually—a view, I may add, that Marx in the closing years of his life actually regarded as possible for Britain and the United States. The Fabians and the early Labour Party believed that socialism could inch its way across Britain through a step-by-step process of political and economic "municipalization," without directly challenging the authority of Parliament and the nation-state. This position was widely held in the formative years of the Labour Party until the party began to gain a respectable number of MPs in the House of Commons, after which its municipal socialism began to take a back seat to typical centralistic demands for the nationalization of industry under top-down boards and managers.

No such socialist or labor movement really developed in the United States. Even the American populists, who gained a mass following in the 1880s and 1890s, directed their efforts primarily toward acquiring power in state legislatures and the federal government. Anarchists, in the name of eschewing all politics as "statist," developed their own indigenous versions of individualism and syndicalism. Although the pre-World War I Socialist Party elected mayors and council members in many American cities, they generally engaged in "sewer socialism," as their limited municipalist approach was to be called, which did little more than create municipally controlled and efficient infrastructures. In this respect, they were not very different from the well-meaning civic reformers in the Democratic and Republican Parties.

But in more positive terms, Americans and Britons share a growing distrust for established national political parties. The "independent voter," as he or she is commonly called, has diminishing, if any, loyalties to the traditional party system. Labour in Britain essentially grows at the expense of Conservative party failures, and the Conservatives at the expense of Labour party failures. Both parties seem to be losing the bedrock support they have historically had: the Conservative-oriented middle classes and the Labour-oriented working classes. Precisely what the future of both national parties will be is very difficult to predict.

The same development is occurring in the United States. Roughly half of the potential American electorate does not vote at all, and the portion that does vote has become overwhelmingly independent in its allegiances of the traditional national parties. In fact, public opinion polls show that most Americans are basically centrist and generally reluctant to give up the social safety net, such as medical care for the elderly, that the Republicans seem intent on shredding. Thus in the United States and Britain alike, the political situation is fluid and potentially volatile.

Of long-range significance is the steady delegitimation that national institutions have undergone in the public mind in both countries. Neither the American Congress nor the British Parliament gain favorable ratings in public opinion polls. A sullenness pervades both countries, a mood of distrust, powerlessness, and cynicism. In the United States the tension between local governments and the increasingly remote national government has grown enormously over recent years. To people in California and New England, as well as in the English Midlands and Scotland, the national center of power seems more distant and out of touch with popular wishes than at any time in the history of this century. The Washington "Beltway" is as remote to a growing number of Americans as Westminster is to many Britons.

In a comprehensive survey of local government in Britain, John Kingdom has summed up the tension between cities and Parliament that could easily apply to the United States. The relationship between municipalities and the nation-state, in his view, is marked by a serious contradiction. By permitting "ordinary people to participate in politics as councilors," the city "plays a major role in legitimation by reducing feelings of powerlessness," he observes. "However, in so doing, local government at the same time threatens the capitalist state. It can be described as its Achilles heel, because it allows ordinary people to be involved in decision making and affords them a moral right to dissent from, and oppose, central government. This means that local government in the modern state is placed in a position of constitutional and political contradiction. The central state both needs local government and fears it," which results in a paradoxical situation whereby the "central government places tension upon local government in almost all its aspects." Although recent Conservative governments in Britain seem to have decided that they can "dispense with the legitimating of local self-government," the fact is that local governments demonstrate "a tenacious power to resist the heavy centralism in the system."[14]

Kingdom's summary was made in 1991, when notions of local control were still in the air and the Greater London Council still remained a fond memory of civic militancy, having been abolished by the Tory government in 1986. What the future holds for British local government is still an open question. In a sharp attack upon the loss of local freedoms, journalist Andrew Rawnsley observes that the Conservative government has "remorselessly stripped local communities of the power to manage their own affairs. Control of local housing, education, transport and health has been transferred to agencies, trusts, boards, development corporations and quangoes, packed with pals of pals, answering not to local electors but a voice in Whitehall whispering instructions down a telephone line."[15]

In a concerted attack upon local government, the Tory government between 1979 and 1995 enacted no less than 150 Acts of Parliament to reduce the powers of town councils and their agencies, shifting some £24 billion in annual spending from elected local authorities to unelected quangos. "The quangocracy cost about one-fifth of government spending in 1992/93," observed a group of journalists for *The Independent on Sunday* and even appointments to these unelected bodies have become less open to public scrutiny and influence.[16] With extraordinary indifference, the Tory government permitted town government to languish, despite many individual protests by local leaders and the liberal press.

Despite the growing encroachment of federal and state governments on local institutions, local government has not declined to this extent in the United States. Although towns and cities, as I have noted, are constitutionally "creatures" of larger statist jurisdictions, American civic institutions still retain a vitality that seems to be waning in present-day Britain. This difference is due in great part to two reasons: the delegitimation of the central government—a process that has probably gone further in the United State than in Britain; and the historical background of how power has been constituted in America.

Although British disenchantment with Westminster is growing as well, the American attitude toward the Washington "Beltway" approximates an active aversion. Indeed, the Republican party's "revolution" in the 1994 election was partly the result of a massive alienation of the electorate from the central government, which seems more remote to Americans than to Britons. The Republicans, exploiting this growing aversion to the "Beltway," called for a devolution of power, mainly from the federal government to state capitals. But it has also awakened mayors and city councils, who have been in a political torpor for decades. City mayors, like state governors, are demanding that their voices be heard by the federal government. The American electorate, in short, seems to have an acerbic attitude toward the central government's concentration of power.

Still another difference has deep historical roots. By the latter half of the nineteenth century, Britain recognized to one degree or another that it was the *urban* "workshop of the world," the center of international industrial capitalism. William Cobbett, in his *Rural Rides*, echoed the dying gasp of the British yeomanry in the 1830s, while Thomas Jefferson, whose life overlapped with Cobbett's, was still preaching the virtues of a mainly agrarian republic. Indeed, it was not until John Steinbeck wrote *The Grapes of Wrath*, his Pulitzer Prize-winning novel of displaced Dust Bowl farmers during the Great Depression, that Americans became fully aware of the fact that their own yeomanry was disappearing.

Yet, despite the massive urbanization of the United States, Americans still tout highly individualistic yeoman values and uphold ideals of self sufficiency that more appropriately belong to a long-gone village life than to a highly industrial and commercial city life. Although any informed Briton knows that the "wild moors" of Bronte novels are manmade and that Sherwood Forest is essentially gone, Americans are still obsessed by myths of a "wilderness" world remote from the artificialities of the city. Cinematic memories of a "Wild West" that was being "tamed" historically by settlers long after Cobbett had gone to his grave, and of "small town" or agrarian virtues not unlike those preached by Jefferson nearly two centuries ago, form a strong part of contemporary mass psychology.

These differences have given to American localism *populist* traits rather than the socialist character it acquired in Britain. It is a populism laced by a Western, or more precisely an acquisitive individualism that, oddly enough, closely conforms to the predatory character of the American bourgeoisie. If a Leftist movement does not arise in the United States that can give the American commitment to small-town virtues a civic agenda that emphasizes libertarian ethics and a communal spirit, there are good reasons to believe that these virtues will be used by the extreme Right to advance a nativist, Christian fundamentalist, and socially reactionary agenda of its own.

Thus, in terms of their similarities, Americans and Britons alike stand at a historic crossroads in which a confederal municipalism offers the only alternative, in my view, to an increasingly authoritarian state or regional coalition of states. Yet in terms of their differences, Americans even more than Britons may fall victim to reactionary ideologies that beggar Thatcherism, Reaganism, and the extremely mean-spirited Republican "revolution" against the poor and underprivileged.

The acquisitive individualism that has been sweeping over the United States has also had an impact on Europe, especially with the "downsizing" of the welfare state in Britain and on parts of the Continent. Today the acquisitive individual constitutes a social malignancy that threatens to destructure and undermine not only social bonds but also the natural world. Its primary effect is simplification, the unraveling of all social ties in the marketplace with its anonymous buyers and sellers, its objectification of all values, its monetization of ideals—and its unrelenting "growth," which turns everything organic into the inorganic, ossifying community and individual alike.

That it may ultimately yield a society divested of nearly all cultural variety, a human psyche divested of nearly all uniqueness, and a natural world divested of nearly all diversity is the intuitive fear that pervades our times, for which nuclear immolation and the wasteland it may yield are as much a metaphor as a real possibility. Thus, politics and citizenship are not the only victims of this corrosive process. If a confederal municipalist orientation can be revived and a sense of citizenship fostered, redolent of the classical meaning of such a perspective, they may well be an antidote for the social pathologies of our time.

But any agenda that tries to restore and amplify the classical meaning of politics and citizenship must clearly indicate what they are *not* if only because of the confusion that surrounds the two words. I cannot repeat often enough that politics is *not* statecraft, and citizens are *not* "constituents" or "taxpayers." Statecraft consists of operations that engage the state: the exercise of its monopoly of violence, its control of the entire regulative apparatus of society in the form of legal and ordinance-making bodies, and its governance of society by means of professional legislators, armies, police forces, and bureaucracies. Statecraft takes on a political patina when so-called "political parties" attempt, in various power plays, to occupy the offices that make state policy and execute it. This kind of "politics" has an almost tedious typicality. A "political party" is normally a structured hierarchy, fleshed out by a membership that functions in a top-down manner. It is a miniature state, and in some countries, such as the former Soviet Union and Nazi Germany, actually constituted the state itself.

The Soviet and Nazi examples of the state *qua* party were the logical extension of the party into the state. Indeed, every party has its roots in the state, not in the citizenry. The conventional party is hitched to the state like a garment to a mannequin. However varied the garment and its design may be, it is not part of the body politic; it merely drapes it. There is nothing authentically political about this phenomenon: it is meant precisely to contain the body politic, to control it and to manipulate it, not to express its will—or even permit it to develop a will. In no sense is a conventional "political" party derivative of the body politic or constituted by it. Leaving metaphors aside, "political" parties are replications of the state when they are out of power and are often synonymous with the state when they are in power. They are formed to mobilize, to command, to acquire power, and to rule. Thus, they are as inorganic as the state itself—an excrescence of society that has no real roots in it, no responsiveness to it beyond the needs of faction, power, and mobilization.

Politics, by contrast, is an organic phenomenon. It is organic in the very real sense that it is the activity of a public body—a community, if you will—just as the process of flowering is an organic activity of a plant. Politics, conceived as an activity, involves rational discourse, public empowerment, the exercise of practical reason, and its realization in a shared, indeed participatory, activity. It is the sphere of societal life beyond the family and the personal needs of the individual that still retains the intimacy, involvement, and sense of responsibility enjoyed in private arenas of life. Groups may form to advance specific political views and programs, but these views and programs are no better than their capacity to answer to the needs of an active public body. A clear failing of many "political" parties is the fact that their programs or "ideologies" are imposed on the public by individuals or their acolytes whose relationship to the community is tenuous and largely conceptual. One thinks here of Karl Marx, whose ideas were developed within the confines of the British Museum and then foisted on the world with a scriptural authority that still generates an endless stream of academic dissertations, even though they exercise virtually no influence in public life.

By contrast, political movements, in the authentic sense of the word, emerge out of the body politic itself, and although their programs are formulated by theorists, they also emerge from the lived experiences and traditions of the public itself. The populist movements that swept out of agrarian America and czarist Russia, or the anarcho-syndicalist and peasant movements of Spain and Mexico articulated deeply felt, albeit often unconscious, public desires and needs. At their best, genuine political movements brought to consciousness the subterranean aspirations of discontented people and eventually turned this consciousness into political cultures that gave coherence to inchoate and formless public desires.

Robert Michels, despite his jaundiced view of the "competence" of the "masses" in *Political Parties* and his proclivity for charismatic leaders, provides a compelling argument for the inertial effect of conventional political parties in periods of rapid social change. They tend to take over institutions that radical movements and the people create rather than innovate them, indeed, ultimately reworking them along statist lines.

The Bolshevik Revolution of 1917–1921 is a textbook example of the appropriation of a popular movement by a highly centralized party. The revolution ended in the evisceration of an elaborate popular council system (the soviets) by a state-oriented party and the complete divestiture of all power that the populace had so painstakingly acquired. A new state apparatus was completely embodied by the Bolshevik Party. By the 1920s in Russia, statecraft had fully replaced politics. And constituents—more properly, subjects of a totalitarian regime—had replaced citizens. The Russian Bolsheviks had introduced a new wrinkle in the concept of a "constituent." Deprived even of any representation in the state, the Russian people were turned by Bolshevism into a "mass." Bolshevism, in effect, established a pattern for "mass mobilization" that was to be emulated and used by National Socialism in Germany.

The recovery and development of politics must, I submit, take its point of departure from the citizen and his or her immediate environment beyond the familial and private arenas of life. There can be no politics without community. And by community I mean a municipal association of people reinforced by its own economic power, its own institutionalization of the grassroots, and the confederal support of nearby communities organized into a territorial network on a local and regional scale. Parties that do not intertwine with these grassroots forms of popular organization are not political in the classical sense of the term. In fact, they are bureaucratic and antithetical to the development of a participatory politics and participating citizens. The authentic unit of political life, in effect, is the municipality, whether as a whole, if it is humanly scaled, or in its various subdivisions, notably the neighborhood.

Nor can politics be structured around the delegation of power. Democracy, conceived as rule by the people, is totally inconsistent with the more republican vision of rule by representatives of the people. It is historical cliché to emphasize that the authors of the United States Constitution, which replaced the Articles of Confederation in the 1780s, meant that *they* and *their* social strata were the "People" in the opening sentence of the document. The Constitution did not create a democracy along Hellenic lines; it created a republic along Roman lines. And like the Roman Republic it unavoidably and reluctantly incorporated inherited democratic institutions such as popular assemblies, which the brothers Gracchi tried to radically expand in the second century B.C.—an endeavor that ended in tragic failure.

All statist objections aside, the problem of restoring municipal assemblies seems formidable if it is cast in strictly structural and spatial terms. New York City and London have no way of "assembling" if they try to emulate ancient Athens, with its comparatively small citizen body. Both cities, in fact, are no longer cities in the classical sense of the term and hardly rate as municipalities even by nineteenth-century standards of urbanism. Viewed in strictly macroscopic terms, they are sprawling urban belts that suck up millions of people daily from communities at a substantial distance from their commercial centers.

But they are also made up of neighborhoods, that is to say, of smaller communities that have a certain measure of identity, whether defined by a shared cultural heritage, economic interests, a commonality of social views, or even an aesthetic tradition such as Greenwich Village in New York or Camden Town in London. However much their administration as logistical, sanitary, and commercial artifacts requires a high degree of coordination by experts and their aides, they are potentially open to political and, in time, physical decentralization. Popular, even block assemblies can be formed irrespective of the size of a city, provided its cultural components are identified and their uniqueness fostered. At the same time, I should emphasize that the confederal municipalist (or equivalently, communalist) views I propound here are meant to be a *changing and formative perspective*—a concept of politics and citizenship to ultimately transform cities and urban megalopolises ethically as well as spatially, and politically as well as economically. Insofar as these views gain public acceptance, they can be expected not only to enlarge their vision and embrace confederations of neighborhoods but also to advance a goal of *physically* decentralizing urban centers. To the extent that mere electoral "constituents" are transformed by education and experience into active citizens, the issue of humanly scaled communities can hardly be avoided as the next step toward a stable and viable form of city life. It would be foolhardy to try to predict in any detail a series of such "next steps" or the pace at which they will occur. Suffice it to say that as a perspective, confederal municipalism is meant to be an ever-developing, creative, and reconstructive agenda as well as an alternative to the centralized nation-state and to an economy based on profit, competition, and mindless growth.

Minimally then, attempts to initiate assemblies can begin with populations that range anywhere from a modest residential neighborhood to a dozen neighborhoods or more. They can be coordinated by strictly mandated delegates who are rotatable, recallable, and above all, rigorously instructed in written form to either support or oppose whatever issue that appears on the agenda of local confederal councils composed of delegates from several neighborhood assemblies. There is no mystery involved in this form of organization. The historical evidence for their efficacy and their continual reappearance in times of rapid social change is considerable and persuasive. The Parisian sections of 1793, despite the size of Paris (between 700,000 and a million inhabitants) and the logistical difficulties of the era (a time when nothing moved faster than a horse), functioned with a great deal of success on their own, coordinated by sectional delegates in the Paris Commune. They were notable not only for their effectiveness in dealing with political issues based on a face-to-face democratic structure, but also for playing a major role in provisioning the city, in preventing the hoarding of food, and in suppressing speculation, supervising the maximum for fixed prices, and carrying out many other complex administrative tasks. Thus, from a minimal standpoint, no city need be considered so large that popular assemblies cannot start, least of all one that has definable neighborhoods that might interlink with each other on ever-broader confederations. The real difficulty is largely administrative: how to provide for the material amenities of city life, support complex logistical and traffic burdens, or maintain a sanitary environment.

This issue is often obscured by a serious confusion between the formulation of policy and its administration. For a community to decide in a participatory manner what specific course of action it should take in dealing with a technical problem does not oblige all its citizens to execute that policy. The decision to build a road, for example, does not mean that everyone must know how to design and construct one. That is a job for engineers, who can offer alternative designs—a very important political function of experts, to be sure, but one whose soundness the people in assembly can be free to decide. To design and construct a road is strictly an administrative responsibility, albeit one that is always open to public scrutiny. If the distinction between policy making and administration is kept dearly in mind, the role of popular assemblies and the people who administer their decisions easily distinguishes logistical problems from political ones, which are ordinarily entangled with each other in discussions on decentralistic politics. Superficially, the assembly system is a "referendum" form of politics: it is based on a "social contract" to share decision making with the population at large, and abide by the rule of the majority in dealing with problems that confront a municipality, a regional confederation of municipalities, or for that matter, a national entity.

Why, then, is there reason to emphasize the assembly form as crucial to self-governance? Is it not enough to use the referendum, as the Swiss do today, and resolve the problem of democratic procedure in a simple and seemingly uncomplicated way? Why can't policy decisions be made electronically at home—as technology enthusiasts have suggested—"autonomous" individuals listening to debates and voting in the privacy of his or her home?

A number of vital issues, involving the nature of citizenship and the recovery of an enhanced classical vision of politics, must be considered in answering these questions. The "autonomous" individual *qua* "voter" who, in liberal theory, forms the irreducible unit of the referendum process is a fiction. Left to his or her own private destiny in the name of "autonomy" and "independence," the individual becomes an isolated being whose very freedom is denuded of the living social and political matrix from which his or her individuality acquires its flesh and blood.

Indeed, "individuality is impaired when each man decides to shift for himself," observes Max Horkheimer in a pithy critique of personalistic atomism, "as the ordinary man withdraws from participation in political affairs, society tends to revert to the law of the jungle, which crushes all vestiges of individuality. The absolutely isolated individual has always been an illusion. The most esteemed personal qualities, such as independence, will to freedom, sympathy, and the sense of justice, are social as well as individual virtues. The fully developed individual is the consummation of a fully developed society. The emancipation of the individual is not an emancipation from society, but the deliverance of society from atomization, an atomization that may reach its peak in periods of collectivization and mass culture."[17]

One can take these observations still further. The dependent individual violates the high premium we place on freedom, will, and the unfettered assertion of ideas. In liberal society, this has led to a mythic individualism that in popular parlance is presumably rugged—that is to say, totally "autonomous," "independent," *and* "self-seeking."

Nevertheless, "rugged individualism" is as little a desideratum as dependence, which we normally associate with a juvenile form of selfhood. The real anthropology of our species involves the prolonged dependence of the infant and young on elders, a socialization process that, until recent times, ultimately led to a deep interdependence in adulthood, not a brash "independence." The notion of independence, which is often confused with independent thinking and freedom, has been so marbled by pure bourgeois egoism that we tend to forget that our individuality depends heavily on community support systems and solidarity. It is not by childishly subordinating ourselves to the community on the one hand or by detaching ourselves from it on the other that we become mature human beings. What distinguishes us as social beings, hopefully with rational institutions, from solitary beings who lack any serious affiliations, is our capacities for solidarity with one another, for mutually enhancing our self-development and creativity, and attaining freedom within a socially creative and institutionally rich collectivity.

"Citizenship" apart from community can be as debasing to our political selfhood as "citizenship" in a totalitarian state. In both cases, we are thrust back to the condition of dependency that characterizes infancy and childhood. We are rendered dangerously vulnerable to manipulation, whether by powerful personalities in private life or by the state and by corporations in economic life. In neither case do we attain individuality or community. Both, in fact, are dissolved by removing the communal ground on which genuine individuality depends. Rather, it is interdependence within an institutionally rich and rounded community—which no electronic media can produce—that fleshes out the individual with the rationality, solidarity, sense of justice, and ultimately the reality of freedom that makes for a creative and concerned citizen.

Paradoxical as it may seem, the authentic elements of a rational and free society are communal, not individual. Conceived in more institutional terms, the municipality is not only the basis for a free society; it is the irreducible ground for genuine individuality as well. The significance of the municipality is all the greater because it constitutes the discursive arena in which people can intellectually and emotionally confront one another, indeed, experience one another through dialogue, body language, personal intimacy, and face-to-face modes of expression in the course of making collective decisions. I speak, here, of the all-important process of *communizing*, of the ongoing intercourse of many levels of life, that makes for *solidarity*, not only the "neighborliness" so indispensable for truly organic interpersonal relationships. The referendum, conducted in the privacy of one's voting booth or in the electronic isolation of one's home, *privatizes* democracy and thereby subverts it. Voting, like registering one's preferences for a particular soap or detergent in an opinion poll, is the total quantification of citizenship, politics, individuality, and the very formation of ideas as a mutually informative process. The mere vote reflects a pre-formulated "percentage" of our perceptions and values, not their full expression. It is the technical debasing of views into mere preferences, of ideals into mere taste, of overall comprehension into quantification such that human aspirations and beliefs can be reduced to numerical digits.

Finally, the "autonomous individual," lacking any community context, support systems, and organic intercourse, is disengaged from the character-building process—the *paideia*—that the Athenians assigned to politics as one of its most important educational functions. True citizenship and politics entail the ongoing formation of personality, education, and a growing sense of public responsibility and commitment that render communizing and an active body politic meaningful, indeed, that give it existential substance. It is not in the privacy of the school, any more than in the privacy of the voting booth, that these vital personal and political attributes are formed. They require a public presence, embodied by vocal and thinking individuals, and a responsive and discursive public sphere, to achieve reality. "Patriotism," as the etymology of the word indicates, is the nation-state's conception of the citizen as a child, the obedient creature of the nation-state conceived as a *paterfamilias* or stern father, who orchestrates belief and commands devotion. To the extent that we are the "sons" and "daughters" of a "fatherland," we place ourselves in an infantile relationship to the state.

Solidarity or *philia*, by contrast, implies a sense of commitment. It is created by knowledge, training, experience, and reason—in short, by a political education developed during the course of political participation. *Philia*, is the result of the educational and self-formative process that *paideia* is meant to achieve. In the absence of a humanly scaled, comprehensible, and institutionally accessible municipality, this all-important function of politics and its embodiment in citizenship are simply impossible to achieve. In the absence of *philia* or the means to create it, we gauge "political involvement" by the "percentage" of "voters" who "participate" in the "political process"—a degradation of words that totally denatures their authentic meaning and eviscerates their ethical content.

In an era of the growing power of nation-states and corporations, when administration, property ownership, production, bureaucracies, and the flow of capital as well as power are notoriously centralized, how can we invoke a localist, municipally-oriented society without seeming to be starry-eyed visionaries? Is this municipalist, decentralist, and participatory vision of self-governance and selfhood utterly incompatible with the overwhelming trend toward public massification? Does the notion of clearly definable, humanly scaled communities not seem to be redolent of atavistic, backward-looking parochial ideas of a pre-modern world, indeed of the "folk community" (*Volksgemeinschaft*) advocated by German Nazism? Do its advocates wish to undo the technological gains achieved by the industrial and technological revolutions of the past two centuries? And how can we change a system based on market competition and profit to one that is oriented toward the satisfaction of human needs?

This work, let me emphasize, is not a handbook for social recipes that meet the taste of every palate. To make it so—and more than enough books exist that profess to do so—is to subvert the very meaning of the confederal municipalist project it seeks to advance by trying to provide a detailed institutional and economic map of what such a future society should or must look like. It is to foist a preconceived grid of ideas on a future that must create itself without being overloaded with blueprints and schemes that reflect the personal proclivities of their authors rather than a rational use of the lessons of the past and present in remaking society. To offer—or to demand—detailed recipes to resolve every problem that every decentralized human community will face is to deny the future any creativity.

But there are certain issues that are often raised that, I believe, deserve serious consideration.

Speaking generally, if we had worked only with given conditions as our parameters, it is doubtful if humanity would have advanced beyond its primate estate in the Pleistocene. Even with all the geographic, climatic, and anatomical odds against them, our ancestors of, say, a million years ago did manage to fashion simple tools and later sophisticate them into remarkably important instruments for acquiring food and creating shelter. In the late years of the twentieth century, this innovative quality has brought our species a degree of technological knowledge that can make it the most creative—and destructive—life form on the planet. Not that all human beings carry this problematic burden on their shoulders. By virtue of their power, the wealthy, who make so many decisions to which the less than wealthy are obliged to defer, carry the greatest burden for the misfortunes that may occur in the years ahead. But the fact remains that a vast reservoir of knowledge is now available to turn the earth into a paradise for ourselves and a fecund environment for nonhuman life forms—or into an indescribable hell.

The problem of dealing with the growing power of nation-states and of centralized corporations, property ownership, production, and the like is *precisely a question of power*—that is to say, who shall have it or who shall be denied any power at all. Michel Foucault has done our age no service by making power an evil as such. Foucauldian postmodernist views notwithstanding, the broad mass of people in the world today lack what they need most, the power to challenge the nation-state and arrest the centralization of economic resources, lest future generations see all the gains of humanity dissipated and freedom disappear from social discourse.

Minimally, if power is to be socially redistributed so that the ordinary people who do the real work of the world can effectively speak back to those who run social and economic affairs, a movement is vitally needed to educate, mobilize, and, using the wisdom of ordinary and extraordinary people alike, *initiate* local steps to regain power in its most popular and democratic forms. Power of this kind must be collected, if we are to take democracy seriously, in newly developed institutions such as assemblies that allow for the direct participation of citizens in public affairs. Without a movement to work toward such a democratic end, including educators who are prepared, in turn, to be educated, and intellectually sophisticated people who can develop and popularize this project, efforts to challenge power as it is now constituted will simply sputter out in escapades, riots, adventures, and protests.

A serious political movement that seeks to advance a confederal municipalist agenda, in turn, must be patient—just as the Russian populists of the nineteenth century were (one of whom is cited in the dedication to this book). The 1960s upsurge, with all its generous ideals, fell apart because young radicals demanded immediate gratification and sensational successes. The protracted efforts that are so direly needed for building a serious movement—perhaps one whose goals cannot be realized within a single lifetime—were woefully absent. Many of the radicals of 50 years ago, burning with fervor for fundamental change, have since withdrawn into the university system they once denounced, the parliamentary positions they formerly disdained, and the business enterprises they furiously attacked.

A confederal municipalist movement, in particular, would not—and should not—achieve sudden success and wide public accolades. The present period of political malaise at best and outright reaction at worst renders any sensational successes impossible. If such a confederal municipalist movement runs candidates for municipal councils with demands for the institution of public assemblies, it will more likely lose electoral races today rather than win even slight successes. Depending upon the political climate at any given time or place, years may pass before it wins even the most modest success.

In any very real sense, however, this protracted development is a desideratum. With rapid success, many naïve members of a municipal electorate expect rapid changes, which no minority, however substantial, can ever hope to achieve at once. For an unpredictable amount of time, electoral activity will primarily be a form of educational activity, an endeavor to enter the public sphere, however small and contained it may be on the local level, and to educate and interact with ever larger numbers of people.

Even where a measure of electoral success on the local level can be achieved, the prospect of implementing a radically democratic policy is likely to be obstructed by the opposition of the nation-state and the weak position of municipalities in modern "democratic" nation-states. Although it is highly doubtful that even civic authorities would allow a neighborhood assembly to acquire the legal power to make civic policy, still less state and national authorities, let me emphasize that assemblies that have no legal power can exercise enormous moral power. A popular assembly that sternly voices its views on many issues can cause considerable disquiet among local authorities and generate a widespread public reaction in its favor over a large region, indeed even on a national scale.

An interesting case in point is the nuclear freeze resolution that was adopted by 88 percent of the 180 town meetings in Vermont in March 1982. Not only did this resolution resonate throughout the entire United States, leading to *ad hoc* "town meetings" in regions of the country that had never seen them, it affected national policy on this issue and culminated in a demonstration of approximately a million people in New York City. Yet none of the town meetings had the "legal" authority to enforce a nuclear freeze, nor did the issue fall within the purview of a typical New England town meeting's agenda.

Historically, in fact, few civic projects that resemble confederal municipalism began with a view to establishing a radical democracy of any sort. The 48 Parisian sections of 1793 that I examined earlier actually derived from the 60 Parisian electoral districts of 1789. These districts were initially established through a complicated process (deliberately designed to exclude the poorer people of Paris) to choose the Parisian members of the Third Estate when the king convoked the Estates General at Versailles. Thereafter the districts, having chosen their deputies, were expected to disband. In fact, the 60 districts refused to desist from meeting regularly, despite their lack of legal status, and a year later became an integral part of the city's government. With the radicalization of the French Revolution, the fearful city and national authorities tried to weaken the power of the districts by reducing their number to 48—hence, the mutation of the old districts into sections. Finally, the sections opened their doors to everyone, some even included women, without any property or status qualifications. This most radical of civic structures, which produced the most democratic assemblies theretofore seen in history, thus slowly elbowed its way into authority, initially without any legal authority whatever and in flat defiance of the nation-state. For all their limitations, the Parisian sections remain an abiding example of how a seemingly non-legal assembly system can be transformed into a network of revolutionary popular institutions around which a new society can be structured.

Conceivably, a centralized state created by a radical party can immediately, if necessary, abolish private property, alter methods of production, and arrest the flow of capital to safe and congenial havens abroad. But it would not be a democracy. Worse still, its power over people, not only its ownership of property and control over the flow of capital, would grow—and, if historical experience is any guide, become all-encompassing. More likely than not, it would increase the structural "complexity" of society by developing a bureaucracy to administer its controls, and, perhaps most importantly, reduce the people to the status of incompetent clients or "constituents." Whatever its economic success might be, the political trajectory it would be obliged to follow might well prove disastrous.

The immediate goal of a confederal municipalist agenda is not to exercise sudden and massive control by representatives and their bureaucratic agents over the existing economy; its immediate goal is to reopen a public sphere in flat opposition to statism, one that allows for maximum democracy in the literal sense of the term, and to create in embryonic form the institutions that can give power to a people generally. If this perspective can be initially achieved only by morally empowered assemblies on a limited scale, at least it will be a form of popular power that can, in time, expand locally and grow over wide regions. That its future is unforeseeable does not alter the fact that its development depends upon the growing consciousness of the people, not upon the growing power of the state—and how that consciousness, concretized in highly democratic institutions, will develop may be an open issue but it will surely be a political adventure.

Be they large or small, such initial assemblies and the movement that seeks to foster them in civic elections remain the only real school for citizenship we have. There is no civic "curriculum" other than a living and creative political realm that can give rise to people who take the management of public affairs seriously. What we must clearly do in an era of commodification, rivalry, anomie, and egoism is to consciously create a public sphere that will inculcate the values of humanism, cooperation, community, and public service in the everyday practice of civic life. Grassroots citizenship goes hand in hand with grassroots politics. The Athenian *polis*, for all its many shortcomings, offers us a remarkable example of how a high sense of citizenship can be reinforced not only by systematic education, but by an etiquette of civic behavior and an artistic culture that adorns ideals of civic service with the realities of civic practice. Deference to opponents in debates, the use of language to achieve consensus, ongoing public discussion in the *agora* in which even the most prominent of the *polis*'s figures were expected to debate public issues with the least known, the use of wealth not only to meet personal needs but to adorn the *polis* itself (thus placing a high premium on the disaccumulation rather than the accumulation of wealth), a multitude of public festivals, dramas, and satires largely centered on civic affairs and the need to foster civic solidarity—all of these and many other aspects of Athens's political culture created the civic solidarity and responsibility that made for actively involved citizens with a deep sense of civic mission.

For our part, we can do no less—and hopefully, in time, considerably more. The development of citizenship must become an art, not merely an education, and a creative art in the aesthetic sense that appeals to the deeply human desire for self-expression in a meaningful political community. It must be a personal art in which every citizen is fully aware of the fact that his or her community entrusts its destiny to his or her moral probity and rationality. If the ideological authority of state power and statecraft today rests on the assumption that the "citizen" is an incompetent being, the municipalist conception of citizenship rests on precisely the opposite. Every citizen would be regarded as competent to participate directly in the "affairs of state"—indeed, what is more important, he or she would be *encouraged* to do so. Every means would be provided, whether aesthetic or institutional, to foster participation in full as an educative and ethical process that turns the citizen's latent competence into an actual reality. Social and political life would be consciously orchestrated to foster a profound sensitivity, indeed an active sense of concern for the adjudication of differences without denying the need for vigorous dispute when it is needed. Public service would be seen as a uniquely human attribute, not a "gift" that a citizen confers on the community or an onerous task that lie or she must fulfill. Cooperation and civic responsibility would become expressions of acts of sociability and *philia*, not ordinances that the citizen is expected to honor in the breach and evade where he or she can do so.

Put bluntly and clearly, the municipality would become a theater in which life in its most meaningful public form is the plot, a political drama whose grandeur imparts nobility and grandeur to the citizenry that forms the cast. By contrast, our modern cities have become in large part agglomerations of bedroom apartments in which men and women spiritually wither away and their personalities become trivialized by the petty concerns of amusement, consumption, and small talk.

The last, and one of the most intractable problems we face, is economic. Today, economic issues tend to center on "who owns what," "who owns more than whom," and, above all, how disparities in wealth are to be reconciled with a sense of civic commonality. Nearly all municipalities have been fragmented by differences in economic status, pitting poor, middle, and wealthy classes against each other often to the ruin of municipal freedom itself as the bloody history of Italy's medieval and Renaissance cities so clearly demonstrates.

These problems have not disappeared in recent times. Indeed, today they are as severe as they have ever been. But what is unique about our own time—a fact so little understood by many liberals and radicals in North America and Europe—is that entirely new *transclass* issues have emerged that concern environment, growth, transportation, cultural degradation, and the quality of urban life in general. Issues that have been produced by urbanization, not by citification. Cutting across conflicting class interests are such transclass issues as the massive dangers of thermonuclear war, growing state authoritarianism, and ultimately global ecological breakdown. To an extent unparalleled in American history, an enormous variety of citizens' groups have brought people of all class backgrounds into common projects around problems, often very local in character, that concern the destiny and welfare of their community as a whole.

Issues such as the siting of nuclear reactors or nuclear waste dumps, the dangers of acid rain, and the presence of toxic dumps—a few of the many problems that beleaguer innumerable municipalities around the world—have united an astonishing variety of people into movements with shared concerns that render a ritualistic class analysis of their motives a matter of secondary importance. Carried still further, the absorption of small communities by larger ones, of cities by urban belts, and urban belts by "standard metropolitan statistical areas" or conurbations has given rise to militant demands for communal integrity and self-government, an issue that surmounts strictly class and economic interests. The literature on the emergence of these transclass movements, so secondary to internecine struggles within cities of earlier times, is so immense that to merely list the sources would require a sizable volume.

I have given this brief overview of an emerging *general social interest* over old particularistic interests to demonstrate that a new politics could easily come into being—indeed, one that would be concerned not only with restructuring the political landscape on a municipal level but the economic landscape as well. The old debates between "private property" and "nationalized property," are becoming threadbare. Not that these different kinds of ownership and the forms of exploitation they imply have disappeared; rather, they are being increasingly overshadowed by new realities and concerns. Private property, in the traditional sense, with its case for perpetuating the citizen as an economically self-sufficient and politically self-empowered individual, is fading away. It is disappearing not because "creeping socialism" is devouring "free enterprise" but because "creeping corporatism" is devouring everyone—ironically, in the name of "free enterprise." The Greek ideal of the politically sovereign citizen who can make a rational judgment in public affairs because he is free from material need or clientage has been reduced to a mockery. The oligarchical character of economic life threatens democracy, such as it is, not only on a national level but also on a municipal level, where it still preserves a certain degree of intimacy and leeway.

We come here to a breakthrough approach to a municipalist economics that innovatively dissolves the mystical aura surrounding corporatized property and nationalized property, indeed workplace elitism and "workplace democracy." I refer to the *municipalization of property*, already raised in the early days of British socialism, as opposed to its corporatization or its nationalization. As for the workplace, public democracy would be substituted for the traditional images of productive management and operation, "economic democracy" and "economic collectivization." Significantly, "economic democracy" in the workplace is no longer incompatible with a corporatized or nationalized economy. Quite to the contrary: the effective use of "workers' participation" in production, even the outright handing over of industrial operations to the workers who perform them, has become another form of time-studied, assembly-line rationalization, another form of the systematic abuse of labor, by bringing labor itself into complicity with its own exploitation.

Many workers, in fact, would like to get away from their workplaces and find more creative types of work, not simply "participate" in planning their own misery. What "economic democracy" meant in its profoundest sense was free, democratic access to the means of life, the guarantee of freedom from material want—not simply the involvement of workers in onerous productive activities that could better be turned over to machines. It is a blatant bourgeois trick, in which many radicals unknowingly participate, that "economic democracy" has been reinterpreted to mean "employee ownership" or that "workplace democracy" has come to mean workers' "participation" in industrial management rather than freedom from the tyranny of the factory, rationalized labor, and "planned production."

A municipal politics, based on communalist principles, scores a significant advance over all of these conceptions by calling for the municipalization of the economy, and its management by the community, as part of a politic of self-management. Syndicalist demands for the "collectivization" of industry and "workers' control" of individual industrial units are based on contractual and exchange relationships between all collectivized enterprises, thereby indirectly reprivatizing the economy and opening it to traditional forms of private property—even if each enterprise is collectively owned. By contrast, confederal municipalism literally *politicizes* the economy by dissolving economic decision making into the civic domain. Neither factory nor land becomes a separate or potentially competitive unit within a seemingly communal collective. Nor do workers, farmers, technicians, engineers, professionals, and the like perpetuate their vocational identities as separate interests that exist apart from the citizen body in face-to-face assemblies. "Property" is integrated into the municipality as the material component of a civic framework, indeed, as part of a larger whole that is controlled by the citizen body in assembly as citizens—not as "workers," "farmers," "professionals," or any other vocationally oriented special-interest groups.

What is equally important, the famous "contradiction" or "antagonism" between town and country, so crucial in social theory and history, is transcended by the township, the traditional New England jurisdiction, in which an urban entity is the nucleus of its agricultural and village environs—not a domineering urban entity that stands opposed to them. A township, in effect, is a small region within still larger ones, such as the county and larger political jurisdictions.

So conceived, the municipalization of the economy should be distinguished not only from corporatization but also from seemingly more "radical" demands such as nationalization and collectivization. Nationalization of the economy invariably has led to bureaucratic and top-down economic control; collectivization, in turn, could easily lead to a privatized economy in a collectivized form with the perpetuation of class or caste identities. By contrast, municipalization would bring the economy *as a whole* into the orbit of the public sphere, where economic policy could be formulated by the *entire* community—notably its citizens in face-to-face relationships working to achieve a general interest that surmounts separate, vocationally defined specific interests. The economy would cease to be *merely* an economy in the conventional sense of the term, composed of capitalistic, nationalized, or "worker-controlled" enterprises. It would become the economy of the *polis* or the municipality. The municipality, more precisely, the citizen body in face-to-face assembly, would absorb the economy into its public business, divesting it of a separate identity that can become privatized into a self-serving enterprise.

What can prevent the municipality, now reinforced by its own economic apparatus, from becoming a parochial city-state of the kind that appeared in the late Middle Ages? Once again I would like to emphasize that anyone who is looking for guaranteed solutions to the problems raised here will not find them in the form of blissfully insulated institutions that take on a life of their own regardless of consciousness and ethics in human affairs.

But if we are looking for countertendencies rather than guarantees, there is an answer that can be given. The most important single factor that led to the weakening of the late medieval city-state was its stratification from within—as a result of differences not only in wealth but in status positions, partly originating in family origins, partly, too, in vocational differences. Indeed, to the extent that the city lost its collective unity and divided its affairs into private and public business, public life itself became segmented into the "blue nails" or plebeians who dyed cloth in cities such as Florence and the more arrogant strata of artisans who produced quality goods. Wealth, too, factored heavily in a privatized economy where material differences could expand and foster a variety of hierarchical differences.

The municipalization of the economy would not only absorb the vocational differences that could militate against a publicly controlled economy; it would also absorb the material means of life into communal forms of distribution. "From each according to his ability and to each according to his needs"—the famous demand of various nineteenth-century socialisms would be institutionalized as part of the public sphere. This traditional maxim, which is meant to assure that people will have access to the means of life irrespective of the work they are capable of performing, would cease to be merely a precarious credo: it would become a practice, a way of functioning politically that is structurally built into the community as a way of existing as a political entity.

Moreover, the enormous growth of the productive forces, rationally and ecologically employed for social rather than private ends, has rendered the age-old problem of material scarcity a moot issue. Potentially, all the basic means for living in comfort and security are available to the populations of the world, notwithstanding the dire—and often fallacious—claims of misanthropes and anti-humanists such as Garrett Hardin, Paul Ehrlich, and regrettably, advocates of "simple living," who can barely be parted from their computers even as they deride technological developments of almost any kind. It is easily forgotten that only a few generations ago, famine was no less a plague than deadly infectious diseases like the Black Death, and that the life expectancy of most people at the turn of the twentieth century in the United States and Europe averaged only 50 years of age.

No community can hope to achieve economic autarchy, nor should it try to do so. Economically, the wide range of resources that are needed to make many of our widely used goods preclude self-enclosed insularity and parochialism. Far from being a liability, this interdependence among communities and regions can well be regarded as an asset—culturally as well as politically. Interdependence among communities is no less important than interdependence among individuals. Divested of the cultural cross-fertilization that is often a product of economic intercourse, the municipality tends to shrink into itself and disappear into its own civic privatism. Shared needs and resources imply the existence of sharing and, with sharing, communication, rejuvenation by new ideas, and a wider social horizon that yields a wider sensibility to new experiences.

The recent emphasis in environmental theory on "self-sufficiency," if it does not mean a greater degree of prudence in dealing with material resources, is regressive. Localism should never be interpreted to mean parochialism; nor should decentralism ever be interpreted to mean that smallness is a virtue in itself. Small is not necessarily beautiful. The concept of *human scale*, by far the more preferable expression for a truly ecological policy, is meant to make it possible for people to completely grasp their political environment, not to parochially bury themselves in it to the exclusion of cultural stimuli from outside their community's boundaries.

Given these coordinates, it is possible to envision a new political culture with a new revival of citizenship, popular civic institutions, a new kind of economy, and a countervailing dual power, confederally networked, that could arrest and hopefully reverse the growing centralization of the state and corporate enterprises. Moreover, it is also possible to envision an eminently practical point of departure for going beyond the town and city as we have known them up to now and for developing future forms of habitation as communities that seek to achieve a new harmonization between people, and between humanity and the natural world. I have emphasized its practicality because it is now clear that any attempt to tailor a human community to a natural "ecosystem" in which it is located cuts completely against the grain of centralized power, be it state or corporate. Centralized power invariably reproduces itself in centralized forms at all levels of social, political, and economic life. It not only is big; it thinks big. Indeed, this way of being and thinking is a condition for its survival, not only its growth.

As for the technological bases for decentralized communities, we are now witnessing a revolution that would have seemed hopelessly utopian only a few decades ago. Until recently, smaller-scale ecotechnologies were used mainly by individuals, and their efficiency barely compared with that of conventional energy sources, such as fossil fuels and nuclear power plants. This situation has changed dramatically in recent years. Around the world, wind turbines have been developed and are currently in use generating electric power at costs that rival and in some cases eclipse power plants fueled by natural gas or coal, not to speak of the additional health and environmental costs of the latter. These comparisons, which can be expected to improve in favor of alternative energy sources in the years to come, have fostered the expansion of nonfossil-fuel sources throughout the entire world. For example, in India as far back as 1994, there was "a major wind boom," according to the Worldwatch Institute.[18]

A similar "boom" has been made in solar power. New solar cells increasingly approximate the costs of conventional energy sources, particularly in heating water for domestic uses. Photovoltaic cells, in which silicon is used to convert solar energy into electrons, have been developed to a point where "thousands of villagers in the developing world [are] using photovoltaic cells to power lights, televisions, and water pumps, needs that are otherwise met with kerosene lamps, lead-acid batteries, or diesel engines." By the mid-1990s more than 200,000 homes in Mexico, Indonesia, South Africa, and some 2,000 in the Dominican Republic had been "solarized," with a good many more to come.[19] This increasingly sophisticated technology, one of the most important—if not the most important—sources of electrical energy in the years to come, is eminently suitable for humanly-scaled communities.

To view technological advances as intrinsically harmful, particularly nonpolluting sources of energy and automated machinery that can free human beings of mindless toil in a rational society, is as shortsighted as it is arrogant. Understandably, people today will not accept a diet of pious moral platitudes that call for "simple means" that presumably will give them "rich ends," whatever these may be, especially if these platitudes are delivered by well-paid academics and privileged Euro-Americans who have no serious quarrel with the present social order apart from whether it affords them access to "wilderness" theme parks.

For the majority of humanity, toil and needless shortages of food are an everyday reality. To expect them to become active citizens in a vital political, ecologically-oriented community while engaging in arduous work for most of their lives, often on empty bellies, is an unfeeling middle-class presumption. Unless they can enjoy a decent sufficiency in the means of life and freedom from mindless, often involuntary toil, it is the height of arrogance to degrade their humanity by calling them "mouths," as many demographers do, or "consumers," as certain very comfortable, environmentalists do.

Indeed, it is the height of elitism and privilege to deny them the opportunity and the *means* for choosing the kind of lifeways they want to pursue. Nor have the well-to-do strata of Euro-American society deprived themselves of that very freedom of choice—a choice, in fact, that they take for granted as a matter of course. Without fostering promising advances in technology that can free humanity as a whole from its subservience to the present, irrational—and, let me emphasize, anti-ecological—social order, we will almost certainly never achieve the free society whose existence is a precondition for harmony between human and human, and between humanity and the natural world.

Which is not to say that we can ignore the need for a visionary ethical ideal. Ironically, it has been the Right's shrewd emphasis on ethics and matters of spirit in an increasingly meaningless world that has given it a considerable edge over the forces of progress. Ironically, Nazism achieved much of its success among the German people a half century ago not because of any economic panaceas it offered, but because of its mythic ideal of nationhood, community, and moral regeneration. In recent times, reactionary movements in America have won millions to their cause on such values as the integrity of the family, religious belief, the renewal of patriotism, and the right to life—a message, I may add, that has been construed not only as a justification for anti-abortion legislation, but as a hypostasization of the individual's sacredness, unborn as well as born. If one thing is clear, it is that just as the Left has emphasized the need to resolve the problems of material scarcity, it is equally necessary to emphasize the need to address the moral emptiness that a market society produces among large numbers of people today.

Morality and ethics, let me add, cannot be reduced to mere rhetoric to match the claims of reactionaries, rather they must be the felt spiritual underpinnings of a new social outlook. They must be viewed not as a patronizing sermon but as a living practice that people can incorporate into their personal lives and their communities. The vacuity and triviality of life today must be replaced precisely by radical ideals of solidarity and freedom that sustain the human side of life as well as its material side, or else the ideals by which a rational future should be guided will disappear in the commodity-oriented world we call the "marketplace of ideas."

The most indecent aspect of this "marketplace" is that ideals tend to become artifacts, mere commodities, that lack even the value of the material things we need to sustain us. They become the ideological ornaments to garnish an inherently antihuman and anti-ecological society, one that threatens to undermine moral integrity as such and the simple social amenities that foster human intercourse.

Thus a municipal agenda that is meant to countervail urbanization and the nation-state must be more than a mere electoral platform, such as we expect from conventional parties. It must also be a message, comparable to the great manifestos advanced by various socialist movements in the last century, which called for moral as well as material and institutional reconstruction. Today's electoral platforms, whether "green" or "red," radical or liberal, are generally shopping lists of demands, precisely suited for that "marketplace of ideas" we have misnamed "politics."

Nor can a municipal agenda be a means for effacing serious differences in outlook. The need for thinking out ideas and struggling vigorously to give them coherence, which alone renders an agenda for a new municipal politics intelligible, is often sacrificed to ideological confusion in the name of achieving a specious "unity." A cranky pluralism is replacing an appreciation of focused thinking; a shallow relativism is replacing a sense of continuity and meaningful values; a confused eclecticism is replacing wholeness, clarity, and consistency. Many promising movements for basic social change in the recent past were plagued by a pluralism in which totally contradictory views were never worked out or followed to their logical conclusions, a problem that has grown even worse today due to the cultural illiteracy that plagues contemporary society.

Finally, at the risk of repetition, it remains to emphasize that a new political agenda can be a municipal agenda only if we are to take our commitments to democracy seriously. Otherwise, we will be entangled with one or another variant of statecraft, a bureaucratic structure that is demonstrably inimical to a vibrant public life. The living cell that forms the basic unit of political life is the municipality, from which everything—such as citizenship, interdependence, confederation, and freedom—emerges. There is no way to piece together any politics unless we begin with its most elementary forms: the villages, towns, neighborhoods, and cities in which people live on the most intimate level of political interdependence beyond private life. It is at this level that they can begin to gain a familiarity with the political process, a process that involves a good deal more than voting and information. It is on this level, too, that they can go beyond the private insularity of family life—a life that is currently celebrated for its inwardness and seclusion—and improvise those public institutions that make for broad community participation and consociation.

In short, it is through the municipality that people can reconstitute themselves from isolated monads into an innovative body politic and create an existentially vital, indeed protoplasmic civic life that has continuity and institutional form as well as civic content. I refer here to the block organizations, neighborhood assemblies, town meetings, civic confederations, and the public arenas for discourse that go beyond such episodic, single-issue demonstrations and campaigns, valuable as they may be, to redress social injustices. Protest alone is not enough; indeed, it is usually defined by what protestors oppose, not by the social changes they may wish to institute. To ignore the irreducible civic unit of politics and democracy is to play chess without a chessboard, for it is on this civic plane that the long-range endeavor of social renewal must eventually be played out.

I have tried in this book to formulate a body of ideas that have meaning for the political restructuring of our times, not another handbook that offers recipes for how to make a "revolution" in one's backyard or on one's front lawn. I have tried to suggest a political philosophy that lends itself to modification, extension, continuity, and a decent regard for the great variety of needs that distinguish one community from another, not a blueprint that dogmatizes and rigidifies the idea of civic freedom into an inflexible credo.

This book has been thoroughly informed by history, not because I seek theoretical precedents, that "legitimate" my views, but rather because the past reveals rational practices that were actually *lived* despite formidable obstacles that stood in their way. That they did not "work" well enough to survive the onslaughts of later, often very irrational developments does not disprove their efficacy. Unpopular as "reason" may be in this postmodernist world, they rest on the logic of democracy, taken in its literal sense of the direct management of social affairs by the people. Anything less, such as representative republics, may be more effective in our "complex" world—a "complexity" that arises largely as a result of the bureaucratic needs of the state and business corporations—but it is not democracy. Indeed, republics like monarchies presuppose that the people are too juvenile or incompetent to run public affairs, which, it must be supposed, can be dealt with only by congressional or parliamentary elites and such sterling executives as Reagan, Bush, and Clinton, or Thatcher and Major.

In any case, political life is not legitimated by how well it "works." If it were, totalitarianism would be easier to legitimate because it is often more "effective" than democracy, just as business techniques are more "effective" than that "wasteful" process that marks debates and free forms of decision making. Doubtless, "time is money," but freedom is a way of life that even money cannot buy—or certainly should not.

My recourse to history has been primarily an endeavor to show that living human beings, not their science-fiction replicas, actually engaged in, and continued to involve themselves in, a political process that may initially seem visionary when it is presented abstractly. I have tried to show that our contemporary market *society*—not only *economy*—is only five or six decades old at best, and our two-and-half-centuries-old market economy was preceded by a far more mixed society and economy, possibly four centuries old, that developed out of a rigidly structured feudal world, later to phase into a chaotically destructured capitalist world.

What men and women have done in the past and, in some degree, continue to do today can certainly be recovered again. And given what we know about their world and the new means at our disposal to improve our own world, we can go much further than they did. But these people were no less real than we are—all rhetoric about the fixity of human nature to the contrary notwithstanding—and what they did in real life centuries ago, we can repeat and do much better.

This book has also been informed by another belief: power that is not retained by the people is power that is given over to the state. Conversely, whatever power the people gain is power that must be taken away from the state. There can be no institutional vacuum where power exists: it is either invested in the people or it is invested in the state. Where the two "share" power, this condition is extremely precarious and often temporary. Sooner or later, the control of society and its destiny will either shift toward the people and their communities at its base or toward the professional practitioners of statecraft at its summit. Only if the whole existing pyramidal social structure is dismembered and radically democratized will the issue of domination as such disappear and be completely replaced by participation and the principle of complementarity. Power, however, must be conceived as real, indeed, solid and tangible, not only as spiritual and psychological. To ignore the fact that power is a muscular fact of life is to drift from the visionary into the ethereal and mislead the public as to its crucial significance in affecting society's destiny.

What this means is that if power is to be regained by the people from the state, the management of society must be deprofessionalized as much as possible. That is to say, it must be simplified and rendered transparent, indeed, clear, accessible, and manageable such that most of its affairs can be run by ordinary citizens. This emphasis on amateurism as distinguished from professionalism is not new. It formed the basis of Athenian democratic practice for generations. Indeed, it was so ably practiced that sortition rather than election formed the basis of the *polis*'s democracy. It resurfaced repeatedly, for example, in early medieval city charters and confederations, and in the great democratic revolutions of the eighteenth century.

Power is also a solid and tangible fact to be reckoned with militarily, notably in the ubiquitous truth that the power of the state or the people eventually reposes in force. Whether the state has power ultimately depends upon whether it exercises a monopoly of violence. Here, too, the Athenian, British, and American yeomen knew only too well that a professional military was a threat to liberty and the state was a vehicle for disarming the people.

A true civicism that tries to create a genuine politics, an empowered citizenry, and a municipalized economy would be a vulnerable project indeed if it failed to replace the police and the professional army with a popular militia—more specifically, a civic guard, composed of rotating patrols for police purposes and well-trained citizen military contingents for dealing with external dangers to freedom. Greek democracy would never have survived the repeated assaults of the Greek aristocracy without its militia of citizen hoplites, those foot soldiers who could answer the call to arms with their own weapons and elected commanders. The tragic history of the state's ascendancy over free municipalities, even the rise of oligarchy within free cities of the past, is the story of armed professionals who commandeered power from unarmed peoples.

Beyond the municipal agenda I have presented thus far lies another, more long-range, agenda: the vision of a political world in which the state as such would finally be replaced completely by a confederal network of municipal assemblies; all socially important forms of property would be absorbed into a truly political economy in which municipalities, interacting with each other economically as well as politically, would resolve their material problems as citizens in open assemblies; and urbanization would give way to humanly scaled and physically decentralized municipalities.

Not only would people then be able to transform themselves from occupational beings into communally-oriented citizens, they would create a world in which all weapons could indeed be beaten into plowshares. Ultimately, it would be possible for new networks of communities to emerge that would be exquisitely tailored—psychologically and spiritually as well as technologically, architecturally, and structurally—to the natural environments in which they exist.

This agenda for a more distant future embodies the "ultimate" vision I have elaborated in greater detail in my previous writings.

Its achievement can no longer be seen as a sudden "revolution" that within a brief span of time will replace the present society with a radically new one. Actually, such revolutions never really happened in history. Even the French Revolution, which radicals have long regarded as a paradigm of sudden social change, was generations in making and did not come to its definitive end until a century later, when the last of the *sans culottes* were exterminated on the barricades of the Paris Commune of 1871.

Nor can we afford today the myth that barricades are more than a symbol. What links my minimal agenda to my ultimate one is a *process*, an admittedly long development in which the existing institutions and traditions of freedom are slowly enlarged and expanded. For the present, we must try increasingly to democratize the republic, a call that consists of preserving—and *expanding*—freedoms we have earned centuries ago, together with the institutions that give them reality. For the future it means that we must radicalize the democracy we create, imparting an even more creative content to the democratic institutions we have rescued and tried to develop.

Admittedly, at that later point we will have moved from a countervailing position that tries to play our democratic institutions against the state into a militant attempt to replace the state with municipally-based confederal structures. It is to be devoutly hoped that by that time, too, the state power itself will have been hollowed out institutionally by local or civic structures, indeed that its very legitimacy, not to speak of its authority as a coercive force, will simply lead to its collapse in any period of confrontation. If the great revolutions of the past provide us with examples of how so major a shift is possible, it would be well to remember that the seemingly all-powerful monarchies which were replaced by republics two centuries ago were so denuded of power that they crumbled rather than "fell," much as a mummified corpse turns to dust after it has been suddenly exposed to air.

Another future prospect also faces us, a chilling one, in which urbanization so completely devours the city and the countryside that community becomes an archaism; in which a market society filters into the most private recesses of our lives as individuals and effaces all sense of personality, let alone individuality; in which a state renders politics and citizenship not only a mockery but a maw that absorbs the very notion of freedom itself.

This prospect is still sufficiently removed from our most immediate experience that its realization can be arrested by those countervailing forces—that dual power—that I have outlined in the previous pages. Given the persistent destructuring of the natural world—of which global warming is only the most threatening example—as well as the social world, more than human freedom is in the balance. The rise of reactionary nationalisms and proliferation of nuclear weapons are but two reminders that we may be reaching a point of cosmic finality in our affairs on the planet. Thus the recovery of a classical concept of politics and citizenship is not only a precondition for a free society; it is also a precondition for our survival as a species. Looming before us is the specter of a disassembled natural world as well as a completely denatured urban world—a natural and social world so divested of its variety and habitability that we will be unable to exist as viable beings.

Confederal Municipalism:
An Overview

Perhaps the greatest single failing of movements for social recon-struction—I refer particularly to the Left, to radical ecology groups, and to organizations that profess to speak for the oppressed—is their lack of a politics that will carry people beyond the limits established by the status quo.

Politics today means duels between top-down bureaucratic parties for electoral office, that offer vacuous programs for "social justice" to attract a nondescript "electorate." Once in office their programs usually turn into a bouquet of "compromises." In this re-spect, many Green parties in Europe have been only marginally dif-ferent from conventional parliamentary parties. Nor have socialist parties, with all their various labels, exhibited any basic differences from their capitalist counterparts. To be sure, the indifference of the Euro-American public—its "apoliticsm"—is understandably de-pressing. Given their low expectations, when people do vote, they normally turn to established parties if only because, as centers of power, they can produce results of sorts in practical matters. If one bothers to vote, most people reason, why waste a vote on a new mar-ginal organization that has all the characteristics of the major ones and that will eventually become corrupted if it succeeds? Witness the German Greens whose internal and public life, by the late 1980s, began to approximate that of other parties in Germany.

That this "political process" has lingered on with almost no basic alternation for decades now is due in great part to the inertia of the process itself. Time wears expectations thin, and hopes are often reduced to habits as one disappointment is followed by another. Talk of a "new politics," of upsetting tradition, which is as old as politics itself, is becoming unconvincing. For decades, at least, the changes that have occurred in radical politics are largely changes in *rhetoric* rather than *structure*. The German Greens are only the latest in a succession of "non-party parties" (to use their original way of describing their organization) that have turned from an attempt to practice grassroots politics—ironically in the Bundestag, of all places!—into a typical parliamentary party. To the modern political imagination, "politics" is precisely a body of *techniques* for holding power in representative bodies—notably the legislative and executive arenas—not a *moral* calling based on rationality, community, and freedom.

A Civic Ethics

Confederal municipalism represents a serious, indeed a historically fundamental project, to render politics ethical in character and grassroots in organization. It is structurally and morally different from other grassroots efforts, not merely rhetorically different. It seeks to reclaim the public sphere for the exercise of authentic citizenship while breaking away from the bleak cycle of parliamentarism and its mystification of the "party" mechanism as a means for public representation. In these respects, confederal municipalism is not merely a "political strategy." It is an effort to work from latent or incipient democratic possibilities toward a radically new configuration of society itself—a communalist society oriented toward meeting human needs, responding to ecological imperatives, and developing a new ethics based on sharing and cooperation. That it involves a consistently independent form of politics is a truism. More important, it involves a *redefinition* of politics, a return to the word's original Greek meaning as the management of the community or *polis* by means of direct face-to-face assemblies of the people in the formulation of public policy and based on an ethics of complementarity and solidarity.

In this respect, confederal municipalism is not one of many pluralistic techniques that is intended to achieve a vague and undefined social goal. Democratic to its core, and nonhierarchical in its structure, it is a kind of human destiny, not merely one of an assortment of political tools or strategies that can be adopted and discarded with the aim of achieving power. Confederal municipalism, in effect, seeks to define the institutional contours of a new society even as it advances the practical message of a radically new politics for our day.

Means and Ends

Here, means and ends meet in a rational unity. The word *politics* now expresses direct popular control of society by its citizens through achieving and sustaining a true democracy in municipal assemblies—this, as distinguished from republican systems of representation that preempt the right of the citizen to formulate community and regional policies. Such politics is radically distinct from statecraft and the State—a professional body composed of bureaucrats, policy, military, legislators, and the like, that exists as a coercive apparatus, clearly distinct from and above the people. The confederal municipalist approach distinguishes statecraft—which we usually characterize as "politics" today—and politics as it once existed in precapitalist democratic communities.

Moreover, confederal municipalism also involves a clear delineation of the social realm—as well as the political realm—in the strict meaning of the term *social*: notably the arena in which we live our private lives and engage in production. As such, the social realm is to be distinguished from both the political and the statist realms. Enormous mischief has been caused by the interchangeable use of these terms—social, political, and the state. Indeed, the tendency has been to identify them with one another in our thinking and in the reality of everyday life. But the State is a completely alien formation, a thorn in the side of human development, an exogenous entity that has incessantly encroached on the social and political realms. Often, in fact, the State has been an end in itself, as witness the rise of Asian empires, ancient imperial Rome, and the totalitarian state of modern times. More than this, it has steadily invaded the political domain, which, for all its past shortcomings, had empowered communities, social groupings, and individuals.

Such invasions have not gone unchallenged. Indeed, the conflict been the State on the one hand and the political social realms on the other has been an ongoing subterranean civil war for centuries. It has often broken out into the open—in modern times in the conflict of the Castilian cities (*Comuneros*) against the Spanish monarchy in the 1520s, in the struggle of the Parisian sections against the centralist Jacobin convention of 1793, and in endless other clashes both before and after these encounters.

Today, with the increasing centralization and concentration of power in the nation-state, a new politics—one that is genuinely new—must be structured institutionally around the restoration of power by municipalities. This is not only necessary but possible even in such gigantic urban areas as New York City, Montreal, London, and Paris. Such urban agglomerations are not, strictly speaking, cities or municipalities in the traditional sense of those terms, despite being designated as such by sociologists. It is only if we think that they *are* cities that we become mystified by problems of size and logistics. Even before we confront the ecological imperative of *physical* decentralization (a necessity anticipated by Friedrich Engels and Peter Kropotkin alike), we need feel no problems about decentralizing them *institutionally*. When François Mitterrand tried to decentralize Paris with local city halls a few years ago, his reasons were strictly tactical (he wanted to weaken the authority of the capital's right-wing mayor). Nonetheless, he failed not because restructuring the large metropolis was impossible, but because the majority of the affluent Parisians supported the mayor.

Clearly, institutional changes do not occur in a social vacuum. Nor do they guarantee that a decentralized municipality, even if it is structurally democratic, will necessarily be humane, rational, and ecological in dealing with public affairs. Confederal municipalism is premised on the struggle to achieve a rational and ecological society, a struggle that depends on education and organization. From the beginning, it presupposes a genuinely democratic desire by people to arrest the growing powers of the nation-state and reclaim them for their community and their region. Unless there is a movement to foster these aims, decentralization can lead to local parochialism as easily as it can lead to ecological humanist communities.

But when have basic social changes ever been without risk? The case that Marx's commitment to a centralized state and planned economy would inevitably yield bureaucratic totalitarianism could have been better made than the case that decentralized libertarian municipalities will inevitably be authoritarian and have exclusionary and parochial traits. Economic interdependence is a fact of life today, and capitalism itself has made parochial autarchies a chimera. While municipalities and regions can seek to attain a considerable measure of self-sufficiency, we have long left the era when self-sufficient communities that can indulge their prejudices are possible.

Confederalism

Equally important is the need for confederation—the interlinking of communities with one another through recallable deputies mandated by municipal citizens' assemblies and whose sole functions are coordinative and administrative. Confederation has a long history of its own that dates back to antiquity, and that surfaced as a major alternative to the nation-state. From the American Revolution through the French Revolution and the Spanish Revolution of 1936, confederalism constituted a major challenge to state centralism. Nor has it disappeared in our own time, when the breakup of existing twentieth-century empires raises the issue of enforced state centralism or the relatively autonomous nation. Confederal municipalism adds a radically democratic dimension to contemporary discussions of confederation by calling for confederations not of nation-states but of *municipalities* and of the neighborhoods of giant megalopolitan areas as well as towns and villages. As I have observed elsewhere in a detailed treatment of this subject, confederalism is thus a way of perpetuating the interdependence that should exist among communities and regions; indeed, it is a way of democratizing that interdependence without surrendering the principle of local control.[1]

In the case of confederal municipalism, parochialism can thus be checked not only by the compelling realities of economic interdependence but by the commitment of municipal minorities to defer to the majority wishes of participating communities. Do these interdependencies and majority decisions guarantee us that a majority decision will be a correct one? Certainly not—but our chances for a rational and ecological society are much better in this approach than in those that ride on centralized entities and bureaucratic apparatuses.

Many arguments against municipalism—even with its strong confederal emphasis—derive from a failure to understand its distinction between policy-making and administration. This distinction is fundamental to confederal municipalism and must always be kept in mind. *Policy* is made by a community or neighborhood assembly of free citizens; *administration* is performed by confederal councils composed of mandated, recallable deputies of wards, towns, and villages. If particular communities or neighborhoods—or a minority grouping of them—choose to go their own way to a point where human rights are violated or where ecological mayhem is permitted, the majority in a local or regional confederation has every right to prevent such malfeasances through its confederal council. This is not a denial of democracy but the assertion of a shared agreement by all to recognize civil rights and maintain the ecological integrity of a region. These rights and needs are not asserted so much by a confederal council as by the majority of the popular assemblies conceived as one large community that expresses its wishes through its confederal deputies. Thus policy-making still remains local, but its administration is vested in the confederal network *as a whole*. The confederation in effect is a Community of communities based on distinct human rights and ecological imperatives.

If confederal municipalism is not to be totally warped of its form and divested of its meaning, it is a desideratum that must be *fought for*. It speaks to a time—hopefully one that will yet come—when people who feel disempowered actively seek empowerment. Existing in growing tension with the nation-state, it is a process as well as a destiny, a struggle to be fulfilled, not a bequest granted by the summits of the state. It is a *dual power* that contests the legitimacy of the existing state power. Such a movement can be expected to begin slowly, perhaps sporadically, in communities here and there that initially may demand only the moral authority to alter the structuring of society before enough interlinked confederations exist to demand the outright institutional power to replace the State. The growing tension created by the emergence of municipal confederations represents a confrontation between the State and the political realms. This confrontation can be resolved only after confederal municipalism forms the new politics of a popular movement and ultimately captures the imagination of millions.

Certain points, however, should be obvious. The people who initially enter into the duel between confederalism and statism will not be the same human beings as those who eventually *achieve* confederal municipalism. The movement that tries to educate them and the struggles that give confederal municipalist principles reality will turn them into active citizens, rather than passive "constituents." No one who participates in a struggle for social restructuring emerges from that struggle with the prejudices, habits, and sensibilities with which he or she entered it. Hopefully, then, such prejudices—like parochialism—will increasingly be replaced by a generous sense of cooperation and a caring sense of interdependence.

Municipalizing the Economy

It remains to emphasize that confederal municipalism is not merely an evocation of all traditional anti-statist notions of politics. Just as it redefines politics to include face-to-face municipal democracies graduated to confederal levels, so it includes a municipalist and confederal approach to economics. Minimally, a confederal municipalist economics calls for the municipalization of the economy, not its centralization into state-owned "nationalized" enterprises on the one hand or its reduction to "worker controlled" forms of collectivistic capitalism on the other. Trade-union control of "worker-controlled" enterprises (that is, syndicalism) has had its day. This should be evident to anyone who examines the bureaucracies that even revolutionary trade unions spawned during the Spanish Civil War of 1936. Today, corporate capitalism too is increasingly eager to bring the worker into complicity with his or her own exploitation by means of "workplace democracy." Nor was the revolution in Spain or in other countries spared the existence of competition among worker-controlled enterprises for raw materials, markets, and profits. Even more recently, many Israeli kibbutzim have been failures as examples of nonexploitative, need-oriented enterprises, despite the high ideals with which they were initially founded.

Confederal municipalism proposes a radically different form of economy—one that is neither nationalized *nor* collectivized according to syndicalist precepts. It proposes that land and enterprises be placed increasingly in the custody of the community—more precisely, the custody of citizens in free assemblies and their deputies in confederal councils. How work should be planned, what technologies should be used, how goods should be distributed are questions that can only be resolved in practice. The maxim "from each according to his or her ability, to each according to his or her needs" would seem a bedrock guide for an economically rational society, provided, to be sure, that goods are of the highest durability and quality, that needs are guided by rational and ecological standards, and that the ancient notions of limit and balance replace the bourgeois marketplace imperative of "grow or die."

In such a municipal economy—confederal, interdependent, and rational by ecological, not simply technological, standards—we would expect that the special interests that divide people today into workers, professionals, managers, and the like would be melded into a general interest in which people see themselves as *citizens* guided strictly by the needs of their community and region rather than by personal proclivities and vocational concerns. Here, citizenship would come into its own, and rational as well as ecological interpretations of the public good would supplant class and hierarchical interests.

This is the moral basis of a moral economy for moral communities. But of overarching importance is the general social interest that potentially underpins all moral communities, an interest that must ultimately cut across class, gender, ethnic, and status lines if humanity is to continue to exist as a viable species.

Some of the opponents of confederal municipalism—and, regrettably, some of its acolytes—misunderstand what confederal municipalism seeks to achieve—indeed, misunderstand its very nature. For some of its instrumental acolytes, confederal municipalism is viewed as a tactical device to gain entry into so-called independent movements and new third parties that call for "grassroots politics." In the name of "confederal municipalism," some acolytes of the view are prepared to blur the tension that they should cultivate between the civic realm and the state—presumably to gain greater public attention in electoral campaigns for gubernatorial, congressional, and other state offices. These people regrettably warp confederal municipalism into a mere "tactic" or "strategy" and drain it of its revolutionary content.

Other heroic individuals who are prepared to do battle (one day) with the cosmic forces of capitalism find that confederal municipalism is too thorny, irrelevant, or vague to deal with and opt for what is basically a form of political particularism. Our spray-can or "alternative cafe" radicals may choose to brush confederal municipalism aside as an impossible tactic but it never ceases to amaze me that well-meaning radicals who are committed to the "overthrow" of capitalism (no less!) find it too difficult to function politically—and, yes, electorally—in their own neighborhoods for a new politics based on a genuine democracy. If they cannot provide a transformative politics for their own neighborhood—a relatively modest task—I find it very hard to believe that they will ever do much harm to the present social system.

Other critics of confederal municipalism dispute the very possibility of a "general interest." If, for such critics, the face-to-face democracy advocated by confederal municipalism and the need to extend the premises of democracy beyond mere justice to complete freedom do not suffice as a "general interest," it would seem to me that the need to repair our relationship with the natural world is certainly a "general interest" that is beyond dispute—and, indeed, it remains the "general interest" advanced by social ecology. It may be possible to co-opt many dissatisfied elements in the present society, but nature is not co-optable. Indeed, the only politics that remains for the Left is one based on the premise that there is a "general interest" in democratizing society and preserving the planet. Now that traditional forces such as the workers' movement have ebbed from the historical scene, it can be said with almost complete certainty that without confederal municipalism the Left will have no politics whatever.

It is critical that we acknowledge this "general interest," created in our time by the very real threat of the ecological devastation of our planet. Capitalism's "grow or die" imperative stands radically at odds with ecology's imperative of interdependence and limit. The two imperatives can no longer coexist with each other—nor can any society founded on the myth that they can be reconciled hope to survive. Either we will establish an ecological society, or society will go under for everyone, irrespective of his or her status.

Will this ecological society be authoritarian, or possibly even totalitarian, a hierarchical dispensation that is implicit in the image of the planet as a "spaceship"? Or will it be democratic? If history is any guide, the development of a democratic ecological society, as distinguished from a command ecological society, must follow its own logic. One cannot resolve this historical dilemma without getting to its roots. Without a searching analysis of our ecological problems and their social sources, the pernicious institutions that we now have will lead to increased centralization and further ecological catastrophe. In a democratic ecological society, those roots are literally the "grass roots" that confederal municipalism seeks to foster.

For those who rightly call for a new technology, new sources of energy, new means of transportation, and new ecological lifeways, can a new society be anything less than a Community of communities based on confederation rather than statism? We already live in a world in which the economy is over-globalized, over-centralized, and over-bureaucratized. Much that can be done locally and regionally is now being done—largely for profit, military needs, and imperial appetites—on a global scale with a seeming complexity that can actually be easily diminished.

If this seems too "utopian" for our time, then so must the present flood of literature that asks for radically sweeping shifts in energy policies, far-reaching reductions in air and water pollution, and the formulation of worldwide plans to arrest global warming and the destruction of the ozone layer be also seen as "utopian." Is it too much, it is fair to ask, to take such demands one step further and call for institutional and economic changes that are no less drastic and that in fact are based on traditions that are deeply sedimented in American—indeed, the world's—noblest democratic and political traditions?

Nor are we obliged to expect these changes to occur immediately. The Left long worked with minimum and maximum programs for change, in which immediate steps that can be taken now were linked by transitional advances and intermediate areas that would eventually yield ultimate goals. Minimal steps that can be taken now include initiating Left Green municipalist movements that propose popular neighborhood and town assemblies—even if they have only moral functions at first—and electing town and city councilors that advance the cause of these assemblies and other popular institutions. These minimal steps can lead step-by-step to the formation of confederal bodies and the increasing legitimation of truly democratic bodies. Civic banks to fund municipal enterprises and land purchases; the fostering of new ecologically-oriented enterprises that are owned by the community; and the creation of grassroots networks in many fields of endeavor and the public weal—all these can be developed at a pace appropriate to changes that are being made in political life.

That capital will likely "migrate" from communities and confederations that are moving toward confederal municipalism is a problem that every community, every nation, whose political life has become radicalized has faced. Capital, in fact, *normally* "migrates" to areas where it can acquire high profits, irrespective of political consider-ations. Overwhelmed by fears of capital migration, a good case could be established for not rocking the political boat at any time. Far more to the point are that municipally-owned enterprises and farms could provide new eco logically valuable and health-nourishing products to a public that is becoming increasingly aware of the low-quality goods and staples that are being foisted on it now.

Confederal municipalism is a politics that can excite the public imagination, appropriate for a movement that is direly in need of a sense of direction and purpose. It offers ideas, ways, and means not only to undo the present social order but to remake it drasti-cally—expanding its residual democratic traditions into a rational and ecological society. Indeed, in my view, confederal municipalism, is precisely the "Commune of communes" for which radicals have fought over the past two centuries. Today, it is the "red button" that must be pushed if a radical movement is to open the door to the pub-lic sphere. To leave that red button untouched and slip back into the worst habits of the post-1968 New Left, when the notion of "power" was divested of utopian or imaginative qualities, is to reduce radical-ism to yet another subculture that will probably live more on heroic memories than on the hopes of a rational future.

Notes

Prologue

1. It is to avoid this loss that I have written a history of popular revolutions, *The Third Revolution*. This vibrant history, its achievements, and its promise must not be lost to future generations.

2. I repeat this point due to the criticism I have received that presumes that I uphold ancient Athens as a "model" or "paradigm." Such criticism obliges me to emphasize that the city I envision as truly rational, free, and ecological has yet to exist and that all my references to cities in history are designed to show that remarkable institutions— and *nothing more than institutions*—that existed in the past deserve our deepest consideration.

3. See in particular my books, *The Ecology of Freedom: The Emergence and Dissolution of Hierarchy* (Oakland: 2005, AK Press; previously Palo Alto: 1984 Cheshire Books) and *The Philosophy of Social Ecology: Essays in Dialectical Naturalism* (Chico: AK Press; new edition forthcoming in 2022).

Chapter One: Urbanization Against Cities

1. George Orwell, *1984* (New York: Signet, 1950), 156.

Chapter Two: From Tribe to City

1. James Mellaart, *Çatalhöyük* (London: Thames and Hudson, 1967), 58.
2. See Murray Bookchin, *The Ecology of Freedom* (Oakland: AK Press, 2005), Chapter 3: The Emergence of Hierarchy.
3. Jean-Jacques Rousseau, *The Social Contract* (New York: Everyman Edition, 1950), 15.

Chapter Three: The Creation of Politics

1. How paradoxical can be judged by the fact that "statified" is a word I have had to create for my own needs to denote the extent of the state's penetration into social and political life. It does not exist in the English language. This term, which I used extensively in my *Post-Scarcity Anarchism,* produced numerous tiffs with a dedicated editor and repeated misprints with a knowledgeable printer.
2. Quite often, in fact, the word *polis,* for which there is no comparable term in English, is translated as *state.*
3. I use the word "civilized" throughout this book to mean literally the world of the *civitas* or city in the broad Latin sense this term was used, not in any culturally pejorative sense. Readers of my book, *The Ecology of Freedom,* will know that the word denotes no monumental advance in the human condition over so-called "primitive" societies—apart from certain technical and scientific amenities that may have lightened humanity's material burdens.
4. Aristotle, *Politics* (London: Loeb Classical Library, 1932), 1326a30-40. Translation modified by author.
5. Ibid., 1280a-1280b.
6. Aristotle, *Politics*, 1252a3. Translation modified by author.
7. Literally, *politea.* The word *republic*—the Latin for *res publica,* literally *public things*—has no meaning or analogue in Greek, and, in this writer's view, no place in the title of Plato's famous dialogue.
8. Lilly Ross Taylor, *Roman Voting Assemblies* (Ann Arbor: University of Michigan Press, 1966), 2.
9. Ibid., 3.
10. Ibid.
11. Jean-Jaques Rousseau, *op. cit.,* 94.
12. Niccolo Machiavelli, *The Prince* (New York: The Modern Library Editions, 1940), 45-46.

Chapter Four: The Ideal of Citizenship

1. Henri Frankfort, *The Birth of Civilization in the Near East* (New York: Doubleday & Co., 1956), 77.
2. M. I. Finley, *Democracy: Ancient and Modern* (New Brunswick, N.J.: Rutgers University Press, 1973), 22.
3. Claude Mossé, *The Ancient World at Work* (New York: W. W. Norton, 1969), 27–28.
4. Alfred Zimmern, *The Greek Commonwealth* (New York: The Modern Library Editions; n.d.), 59.
5. Plutarch, "Solon" in *The Rise and Fall of Athens* (Harmondsworth: Penguin Books, 1960), 54.
6. Ibid., 62.
7. Quoted in Thucydides, *The Peloponnesian War* (New York: Modern Library Editions, 1944), 121–22.
8. T. B. L. Webster, *Life in Classical Athens* (London: B. T. Batsford, Ltd., 1969), 87.
9. W. G. Forrest, *The Emergence of Greek Democracy* (New York: McGraw-Hill Book Co., 1966), 214.
10. Aeschylus, *Oresteia* (Chicago: University of Chicago Press, 1953), 735–40.
11. Ibid., 681–706.
12. Forrest, *The Emergence of Greek Democracy*, 204; George Thomson, *Aeschylus and Athens* (New York: Grosset & Dunlop, 1968).
13. M. Rostovtzeff, *Rome* (London: Oxford University Press, 1960), 104.
14. Ibid., 100.
15. Ibid., 104.
16. Heinrich Heine, *Reisebilder*, quoted by Ian Scott-Kilvert in *Makers of Rome: Nine Lives by Plutarch* (Harmondsworth: Penguin, 1965), 12.

Chapter Five: Patterns of Civic Freedom

1. Chester Starr, *Civilization and the Caesars* (New York: W. W. Norton, 1965), 90.
2. Ibid., 105.
3. For a more detailed discussion of the decline of Rome and the problems the empire created, the reader should see my book, *The Limits of the City* (New York: Harper Colophon Books, Harper & Row, 1974), 32–35.
4. John H. Mundy and Peter Riesenberg, *The Medieval Town* (New York: Van Nostrand Reinhold Co., 1958), 18.
5. Lauro Martines, *Power and Imagination* (New York: Vintage Books, 1980).

6. John H. Mundy in "Introduction" to Henri Pirenne, *Early Democracy in the Low Countries* (New York: W. W. Norton, 1963), xxii fn.

7. Martines, *Power and Imagination*, 27.

8. John H. Mundy, *Europe in the High Middle Ages* (London: Longman Group Ltd., 1973), 409.

9. Martines, *Power and Imagination*, 35–36.

10. This is a feature that existed even in New York during the 1930s, which I distinctly recall as a young man. This great city, at one time an agglomeration of a thousand ethnically unique neighborhoods, seemed to have been stopped in time, for a generation or two, and created much the same degree of neighborhood loyalties and distinctive accents. I have encountered such differences in New Orleans, although they are rapidly disappearing. The emergence of civic variety, loyalty, and cultural diversity even in modern American cities is an important theme to which we shall return later in this book, and the consequences of its loss will be fully explored as a consequence of contemporary urbanization. Quote is from Martines, *Power and Imagination*, 37.

11. Daniel Waley, *The Italian City Republics* (New York: McGraw-Hill Book Co., 1969), 63.

12. Ibid., 63.

13. Martines, *Power and Imagination*, 49.

14. Ibid., 52.

15. Quoted in John H. Mundy, *Europe in the High Middle Ages*, 408.

16. Ephraim Emerton, *The Beginnings of Modern Europe* (New York: Ginn and Co., 1917), 207.

17. Waley, *The Italian City Republics*, 221.

18. Ibid.

19. Benjamin Barber, *The Death of Communal Liberty* (Princeton: Princeton University Press, 1974), 263.

20. France, it is worth noting, had extended its newly adopted metric system from weights, measures, lengths, and areas directly into the calendar with the result that a week was measured in ten days. This did not displease the French bourgeoisie, which was only too glad to have a longer work week than existed under the older religious calendar.

21. F. Furet, C. Mazauric, and L. Bergeron, "The Sans-Culottes and the French Revolution" in *New Perspectives on the French Revolution*, Jeffry Kaplow, ed. (New York: John Wiley & Sons, 1965), 235.

22. Ibid., 234–35.

23. Cited in Daniel Guérin, *Class Struggles in the French Revolution* (London: Pluto Press, 1977), 32–33.

Chapter Six: From Politics to Statecraft

1. R. R. Palmer, *The Age of Democratic Revolutions* (Princeton: Princeton University Press, 1959).

2. Jacob Burckhardt, *The Civilization of the Renaissance in Italy* (New York: Phaidon Publishers, 1950), 2.

3. J. A. O. Larsen, *Greek Federal States* (London: Oxford University Press, 1967), 27.

4. Peter Kropotkin, *Mutual Aid* (Montreal: Black Rose Books, n.d.).

5. Ibid., 204–205.

6. Ibid., 205.

7. F. Grenfell Baker, *The Model Republic* (New York, 1892), 308. Quoted in Barber, *The Death of Communal Liberty*, 14–15.

8. Lewis Mumford, *The City in History* (New York: Harcourt Brace and World, 1961), 339–40.

9. Waley, *The Italian City Republics*, 126.

10. Hence my strong objections to the way European, particularly German, city confederacies are treated in the mainstream historical literature of the time, particularly by Toynbee and Mumford. In this respect, Kropotkin's writings are still exceptional for their sympathy, although they are not given sufficient attention in his work.

11. Perez Zagorin, *Rebels and Rulers: 1500–1600*, Vol. I (New York: Cambridge University Press, 1982), 232.

12. Friedrich Engels, *The Peasant War in Germany* (New York: International Publishers, 1926), 150.

Chapter Seven: The Social Ecology of Urbanization

1. Cf. Joseph R. Strayer, *On the Medieval Origins of the Modern State* (Princeton: Princeton University Press, 1970).

2. Cf. Eric Hobsbawm, *Primitive Rebels* (Manchester: Manchester University Press, 1959).

3. Perez Zagorin, *Rebels and Rulers*, Vol. 1, 93.

4. Karl Marx, "Preface" to *A Contribution to the Critique of Political Economy Selected Works*, Vol. I (Moscow: Progress Publishers, 1969), 504.

5. Perez Zagorin, *Rebels and Rulers*, 243.

6. Ibid., 244.

7. Mary Beard, *A History of Business*, Vol. I (Ann Arbor: University of Michigan Press, 1938), 50.

8. R. S. Lopez, "The Evolution of Land Transport in the Middle Ages," *Past and Present* (April 1956): 17.

9. Ibid., 18.

10. Ibid., 17.

11. Which is not to say that grain, that all-important import of the Greek cities and Rome, was not a major staple in the Mediterranean basin and along European water routes. But the inland transportation of grain by carts was quite secondary to luxury objects once we look beyond a regional economy in Europe.

12. I find it fascinating that so many radicals, socialists and anarchists alike, almost intuitively rally to the support of small-scale farmers against agribusiness and small family enterprises against giant supermarkets, this despite their hostility to private property of any kind. Americans seem to suspect that domestic forms of society, however propertied, are a desideratum in themselves and foster individuality in a mass society. This Jeffersonian legacy acts as one of the most important inertial elements against the full colonization of society by corporations and big business, even though its historical context in the making of the free city and free citizen has not been fully understood; so, too, for our appreciation of craftsmanship.

13. Cf. Paul Sweezy, "A Critique" and "A Rejoinder" in *The Transition From Feudalism to Capitalism*, Rodney Hilton, ed. (London: New Left Books, 1976), 33–56, 102–108.

14. Put bluntly: there was no "transition" from feudalism to capitalism in the simple sense presented by most Marxist and liberal historians and theorists. Nor is it true to say that capitalism developed "within the womb" of feudalism and then "emerged" as a dominant social order through a series of "bourgeois-democratic revolutions" whose "paradigm" is the French Revolution of 1789–94 (in the view of some historians, the span of the revolution is expanded to include the Napoleonic era, notably up to 1814!). The remarkably mixed, diversified, and complex society that is often presented to us as a "transitional period" was, in fact, an era in its own right that social ecology can describe and interpret more insightfully than economistic interpretations. We shall see that the view of this remarkable era as merely "transitional" leads to a host of errors, particularly its depiction of the revolutionary era and its democratic aspirations as "bourgeois," an imagery that makes capitalism a system more committed to freedom, or even ordinary civil liberties, than it was historically.

15. Cf. Immanuel Wallerstein, *Historical Capitalism* (London: Verso Editions, 1983).

16. This I believe probably accounts for Karl Marx's strongly productivist bias in economic theory over the more consumptionist biases of contemporary economists. Marx, I believe, was quite correct when he pointed out that under capitalism, at least, it was production that created demand, not demand that created production, although there is

an obvious interplay between the two at a surface level of economic life. Where he was most lacking was in his failure to recognize the cultural constraints that limited production and profoundly influenced the vulnerability of a society to technological innovation. In this respect, ancient society was always a mystery to him, and its development was often explained in surprisingly conventional ways.

17. Friedrich Engels, "The Condition of the Working Class in England in 1844," in *Marx-Engels Collected Works*, Vol. 4 (New York: International Publishers, 1975), 320.

18. Ibid., 308.

19. That Engels's commitment to a harsh concept of technological progress, indeed, the whole Marxist theory of "historical materialism," was meant to slap not only the face of philosophical idealism but also of the high spirit of European romanticism is an issue that has yet to be fully explored. Gray became a favorite color of Marxian socialism as part of its deliberate endeavor to disenchant the world and relegate the past, with all its humane as well as barbarous traditions, to a historical "junk heap." Eventually socialism, like Puritanism, was to be deprived of all its ethical content and turned into a doctrine very much like the egoistic political economy of the industrial bourgeoisie itself.

20. Ibid., 398.

21. Ibid., 319.

22. Ed. note: And for many people, social media has now replaced them all. Here, as in the paragraph above, Bookchin foreshadows the enormous impact of social media and its algorithms, which have resulted in sweeping changes in what news people view and what information they see.

Chapter Eight: The New Municipal Agenda

1. Jane Jacobs, *Cities and the Wealth of Nations*, (New York: Random House; 1984).

2. H. Mooseburger, *Die Bundnerische Allemand* (1981), 5. Quoted in Benjamin Barber, op. cit., 112.

3. Benjamin Barber, ibid., 15.

4. Ibid., 49.

5. Ibid., 100.

6. Herman Weilemann: Pax Helvetica: *Oder Die Demokratie der Kleinen Gurppen* (Zurich: 1951). Quoted in Benjamin Barber, ibid., 101.

7. T. H. Breen: *Puritans and Adventurers* (New York: Oxford University Press; 1980), 3.

8. Ibid., 4-5.

9. *Massachusetts Centinel*, June 24, 1786. Quoted in David P. Szatmary: *Shay's Rebellion* (Amherst: University of Massachusetts; 1980), 1.

10. George Richards Minot, quoted in Jackson Turner Main: *Political Parties Before the Constitution* (New York: Norton; 1974), 96 fn.

11. David P. Szatmary, *op. cit.,* 6-7.

12. Robert A. Gross: *The Minute Men and Their World* (New York: Hill and Wang; 1976), 10-11.

13. H. N. Brailsford: *The Levellers in the English Revolution* (Nottingham: Spokesman; 1976), 376.

14. John Kingdom: *Local Government and Politics in Britain* (London and New York: Philip Allan; 1991), 18.

15. Andrew Rawnsley: "Democracy's Brave New Yawn," *The Observer* (May 1, 1994).

16. Nick Cohen, Judith Judd, Judy Jones, and Barrie Clement, "What Happened to Democracy?," *The Independent on Sunday,* (March 28 1993).

17. Max Horkheimer: *The Eclipse of Reason* (New York: Oxford University Press; 1947), 135.

18. Lester Brown *et al: State of the World 1995* (New York and London: W. W. Norton, 1995), 60-70.

19. Ibid., 67.

Appendix: Confederal Municipalism: An Overview

1. See the essay, "The Meaning of Confederalism" in *The Next Revolution: Popular Assemblies and the Promise of Direct Democracy,* (New York and London: Verso Books, 2015).

Index

A

absolutism: confederacies against, 153–154; in English monarchy, 146–147, 194; in Europe, 190–192; in France, 149–150; nation-building during era of, 191–200; in Spain, 151–152; Spanish movement against, 171–176

Achaean League, 154, 155, 158

Aeskylos (trilogy), 78–80

agora, 23; citizens daily life in, . 62–64

agrarian regions: civic freedom in, 91–100. *See also* Gracchi brothers; peasant revolts

agrarian values: during the Depression, 224–227, 253

agribusiness: earliest example of, 154; modern, 129, 204n

"American Dream," 220, 237

American Revolution, 86, 119, 243, 245, 246, 291

ancient cities, 17–31, 95–96; Çatalhöyük, 19–22, 25–27;

dependence on agrarian communities, 91–94. *See also* Athens (Greece); Roman Empire

Antigone, 80–81

Arendt, Hannah, 34, 56, 152

arete, 61, 75

Aristotle, 37–41, 66; on *plebs* vs. *populus,* 115; on population vs. self-sufficiency, 37–38; sociality as civic attribute, 39–40

artisans. *See* guilds

assemblies: in Attica, 73–74; confederal municipalist, 265–270; in Greek confederacy, 158; Greek *ekklesia* (popular assembly), 45, 52, 54, 55, 62, 74, 88; Italian popular *(conjuratio)*, 103–105; municipal, 244, 258–260; New England town meeting, 42, 119, 120, 237, 242; political, 60; Roman, 43–46; sectional (French Revolution), 120–127, 171, 200, 219, 259; tribal, 29–30, 59–60

AK PRESS is small, in terms of staff and resources, but we also manage to be one of the world's most productive anarchist publishing houses. We publish close to twenty books every year, and distribute thousands of other titles published by like-minded independent presses and projects from around the globe. We're entirely worker run and democratically managed. We operate without a corporate structure—no boss, no managers, no bullshit.

The **FRIENDS OF AK PRESS** program is a way you can directly contribute to the continued existence of AK Press, and ensure that we're able to keep publishing books like this one! Friends pay $25 a month directly into our publishing account ($30 for Canada, $35 for international), and receive a copy of every book AK Press publishes for the duration of their membership! Friends also receive a discount on anything they order from our website or buy at a table: 50% on AK titles, and 30% on everything else. We have a Friends of AK ebook program as well: $15 a month gets you an electronic copy of every book we publish for the duration of your membership. *You can even sponsor a very discounted membership for someone in prison.*

Email **friendsofak@akpress.org** for more info, or visit the website: **https://www.akpress.org/friends.html**.

There are always great book projects in the works—so sign up now to become a Friend of AK Press, and let the presses roll!

Murray Bookchin

One of the most important radical thinkers of the last century, Murray Bookchin (1921–2006) originated a reconstructive social theory called "social ecology," blending aspects of classical Greek and modern philosophy, anarchism, anthropology, and ecology in an effort to rethink humanity's relationship with nature. His groundbreaking essay, "Ecology and Revolutionary Thought" (1964), was one of the first to assert that capitalism's grow-or-die ethos was on a dangerous collision course with the natural world that would include the devastation of the planet by global warming. A long-time activist, and author of two dozen books on ecology, history, philosophy, and urbanization, Bookchin insisted that a complete transformation in social relations, in which all forms of hierarchy and domination were eliminated, was essential if we are to heal our relationship with nature. His work has influenced numerous movements around the world, including the New Left of the 1960s, the alterglobalization movement, the radical municipalism movement, and the Kurdish democratic confederalism project in Turkey and Northeast Syria.